Virtual Reality:
Recent Advancements,
Applications and Challenges

RIVER PUBLISHERS SERIES IN AUTOMATION, CONTROL AND ROBOTICS

Series Editors:

ISHWAR K. SETHI
Oakland University
USA

TAREK SOBH
University of Bridgeport
USA

QUAN MIN ZHU
University of the West of England
UK

Indexing: All books published in this series are submitted to the Web of Science Book Citation Index (BkCI), to SCOPUS, to CrossRef and to Google Scholar for evaluation and indexing.

The "River Publishers Series in Automation, Control and Robotics" is a series of comprehensive academic and professional books which focus on the theory and applications of automation, control and robotics. The series focuses on topics ranging from the theory and use of control systems, automation engineering, robotics and intelligent machines.

Books published in the series include research monographs, edited volumes, handbooks and textbooks. The books provide professionals, researchers, educators, and advanced students in the field with an invaluable insight into the latest research and developments.

Topics covered in the series include, but are by no means restricted to the following:

- Robots and Intelligent Machines
- Robotics
- Control Systems
- Control Theory
- Automation Engineering

For a list of other books in this series, visit www.riverpublishers.com

Virtual Reality: Recent Advancements, Applications and Challenges

Editors

Lila Bozgeyikli

School of Information, University of Arizona
USA

Ren Bozgeyikli

School of Information, University of Arizona
USA

River Publishers

Published, sold and distributed by:
River Publishers
Alsbjergvej 10
9260 Gistrup
Denmark

www.riverpublishers.com

ISBN: 978-87-7022-142-9 (Hardback)
      978-87-7022-141-2 (Ebook)

©2020 River Publishers

Contents

Preface

This book is motivated by the rapid advancement in virtual reality (VR) technologies and the wide array of application areas with different implications on user experience. This book differs from the other books in this area with the recentness of its content. Virtual reality technologies advance with a rapid rate; hence studies have severe implications on the future of this technology. With this motivation, in this book, only recent academic studies were included with the hopes of leveraging future research and industry applications through a deeper understanding of application areas, use cases, user experience, metrics, and inspiration by the discussed future research directions.

The covered topics include recent application areas of virtual reality, user experience measurement in virtual reality, intersection of virtual reality and other widely adopted domains, such as robotics in virtual reality. Challenges in the area of virtual reality along with future directions are discussed. A snapshot of the included chapters are as follows:

Chapter 1 begins with a snapshot of recent application areas of virtual reality, such as education, training, health and well-being, and entertainment. Novel interaction techniques and user interfaces are also presented and discussed.

Chapter 2 includes a discussion of digital and visual literacy, video games, and virtual reality. The aim of this chapter is to give a deeper understanding of gamified and immersive learning in today's world as well as discussing the capability of video games and virtual reality to enable experiences that had been previously thought as impossible.

Chapter 3 reviews virtual reality in movement disorders by establishing and understanding the underlying fluctuations relative to anxiety in human postural control when they are exposed to immersive virtual conditions. Effectiveness of virtual reality for evaluating the effects of fear of falling on balance and gait function is discussed. The chapter concludes with the examination of the potential of virtual reality in neurorehabilitation that is aimed at ameliorating fall-related anxiety in adults.

Chapter 4 includes an overview of the utilization of virtual reality in robotics applications with several recent research studies, such as body-in-the-loop control of soft robotic exoskeletons during virtual manual labor tasks, drone positioning training, robot programming, robot trajectory programming and execution, shared control handheld robotic systems in augmented reality, and smart devices, such as connected pianos that are controlled with augmented reality.

Chapter 5 discusses several case studies in developing composition software and media systems for enactive steering of computational models with the use of an enactivist framework to guide the development of simulations in this area. In particular, this chapter includes a discussion of a media system known as EMA (i.e., An Experiential Model of the Atmosphere), and the media composition framework used in its development (i.e., SC).

Chapter 6 provides an overview of recent studies aiming to improve user experience in virtual reality through novel use-cases, interaction techniques and user interfaces, and models of performance assessment. The chapter concludes with a discussion of the emerging challenges for the field of human-computer interaction with a focus on the use of virtual reality in education and training.

The book is addressing industry professionals and university audiences (both researchers and graduate-level college students) who are interested in the field of virtual reality.

We are grateful for the valuable contributions of the authors and the publishers and we hope that this collection of recent studies in virtual reality helps with the advancement of this rapidly evolving field.

Lila Bozgeyikli, PhD
Evren Bozgeyikli, PhD
School of Information, University of Arizona, USA, 2019

Acknowledgments

We would like to thank all the authors for their efforts in contributing to this book in their busy schedules. We would like to also thank River Publishers, especially Mark de Jongh, who gave us the idea of starting this book project and encouraged and helped us throughout the whole process; Rajeev Prasad who helped us with the completion and publishing of the book; and Junko Nakajima, who helped with the production of the book.

The list of authors who contributed to this book in alphabetical order is as follows: Adekunle Adeniyi Ajibade, Annika Wollschläger, Brandon Mechtley, Camilo Perez Quintero, Christopher Roberts, Chung Hyuk Park, Daniel M. Lofaro, Edgar Endress, Elizabeth Croft, Elizabeth T. Hsiao-Wecksler, Fotis Liarokapis, Frank Lee, Franz Steinmetz, German Espinosa, Girish Krishnan, H. F. Machiel Van der Loos, Jack Clark, Joshua Elsdon, Julian Stein, Katey Corbett, Limin Zhang, Manuel E. Hernandez, Maram Sakr, Matthew K. X. J. Pan, Michael Rubenstein, Mike Brayshaw, Neil A. Gordon, Nicholas Thompson, Panagiotis Petridis, Paul Mattioli, Rachneet Kaur, Redwan Alqasemi, Richard Sowers, Rubein Shaikh, Sha Xin Wei, Todd Ingalls, Victoria Uren, Wesley P. Chan, Yafei Xu, Yiannis Demiris.

List of Contributors

Adekunle Adeniyi Ajibade, *Coventry University, UK*

Annika Wollschläger, *Franka Emika GmbH, Germany*

Brandon Mechtley, *School of Arts, Media, and Engineering, Arizona State University, USA; E-mail: bmechtley@asu.edu*

Camilo Perez Quintero, *Collaborative Advanced Robotics and Intelligent Systems (CARIS) Laboratory, Mechanical Engineering University of British Columbia, Canada*

Christopher Roberts, *School of Arts, Media, and Engineering, Arizona State University, USA*

Chung Hyuk Park, *Department of Biomedical Engineering, School of Engineering and Applied Science, George Washington University, USA*

Daniel M. Lofaro, *Department of Electrical and Computer Engineering, George Mason University, USA*

Edgar Endress, *School of Art, George Mason University, USA*

Elizabeth Croft, *Monash University, Australia*
Franka Emika GmbH, Germany

Elizabeth T. Hsiao-Wecksler, *Department of Mechanical Science and Engineering, University of Illinois at Urbana-Champaign, USA*

Evren Bozgeyikli, *School of Information, University of Arizona, USA; E-mail: rboz@email.arizona.edu*

Fotis Liarokapis, *Masaryk University, Czech Republic*

Frank Lee, *College in the Media Arts and Design, Drexel University, USA*

Franz Steinmetz, *German Aerospace Center (DLR) Institute of Robotics and Mechatronics, Germany*

German Espinosa, *Electrical Engineering and Computer Science Department, Northwestern University, USA*

Girish Krishnan, *Department of Industrial and Enterprise Systems Engineering, University of Illinois at Urbana-Champaign, USA*

H. F. Machiel Van der Loos, *Collaborative Advanced Robotics and Intelligent Systems (CARIS) Laboratory, Mechanical Engineering University of British Columbia, Canada*

Jack Clark, *School of Information, University of Arizona, USA; E-mail: jaclarkent@email.arizona.edu*

Joshua Elsdon, *Personal Robotics Laboratory, Imperial College, London, UK*

Julian Stein, *School of Arts, Media, and Engineering, Arizona State University, USA*

Katey Corbett, *Center for Assistive, Rehabilitation and Robotics Technologies, University of South Florida, USA*

Lila Bozgeyikli, *School of Information, University of Arizona, USA; E-mail: lboz@email.arizona.edu*

Limin Zhang, *School of Information, University of Arizona, USA*

Manuel E. Hernandez, *Department of Kinesiology and Community Health, University of Illinois at Urbana-Champaign, USA; E-mail: mhernand@illinois.edu*

Maram Sakr, *Collaborative Advanced Robotics and Intelligent Systems (CARIS) Laboratory, Mechanical Engineering University of British Columbia, Canada*

Matthew K. X. J. Pan, *Collaborative Advanced Robotics and Intelligent Systems (CARIS) Laboratory, Mechanical Engineering University of British Columbia, Canada*

Michael Rubenstein, *Electrical Engineering and Computer Science Department, Northwestern University, USA*

Mike Brayshaw, *Department of Computer Science and Technology, University of Hull, UK*

Neil A. Gordon, *Department of Computer Science and Technology, University of Hull, UK*

Nicholas Thompson, *Department of Industrial and Enterprise Systems Engineering, University of Illinois at Urbana-Champaign, USA*

Panagiotis Petridis, *Aston University, UK*

Paul Mattioli, *Center for Assistive, Rehabilitation and Robotics Technologies, University of South Florida, USA*

Rachneet Kaur, *Department of Industrial and Enterprise Systems Engineering, University of Illinois at Urbana-Champaign, USA; E-mail: kaurrachneet6@gmail.com*

Redwan Alqasemi, *Center for Assistive, Rehabilitation and Robotics Technologies, University of South Florida, USA*

Richard Sowers, *Department of Industrial and Enterprise Systems Engineering, Department of Mathematics, University of Illinois at Urbana-Champaign, USA*

Rubein Shaikh, *Center for Assistive, Rehabilitation and Robotics Technologies, University of South Florida, USA*

Sha Xin Wei, *School of Arts, Media, and Engineering, Arizona State University, USA*

Todd Ingalls, *School of Arts, Media, and Engineering, Arizona State University, USA*

Victoria Uren, *Aston University, UK*

Wesley P. Chan, *Collaborative Advanced Robotics and Intelligent Systems (CARIS) Laboratory, Mechanical Engineering University of British Columbia, Canada*

Yafei Xu, *School of Information, University of Arizona, USA*

Yiannis Demiris, *Personal Robotics Laboratory, Imperial College, London, UK*

List of Figures

List of Tables

List of Abbreviations

3D	Three-dimensional
A-CBT	Additional cognitive behavioral therapy
ANDI	Autism & Neurodevelopmental Disorders Institute at George Washington University in Washington D.C.
ANOVA	analyses of variance
AQ	Acrophobia questionnaire
AR	Augmented reality
ASD	Autism spectrum disorder
BCI	Brain-computer interface
BED	Binge eating disorder
BN	Bulimia nervosa
CAVE	Cave automatic virtual environment
CBT	Cognitive behavioral therapy
CCI	Co-contraction index
CET	Cue exposure therapy
CGI	Computer-generated imaging
CNN	Convolution neural network
DLR	German Aerospace Center
DOF	Degree of freedom
EAI	Everyday action impairment
EDA	Electrodermal activity
EDI-3	Eating disorder inventory 3
EEG	Electroencephalogram
EKG	Electrocardiography
EMG	Electromyography
ERP	Event-related potential
FAA	Frontal alpha asymmetry index
FCQ-T/S	The state and trait food craving questionnaire
FIR	Finite impulse response
fNIRS	Functional near-infrared spectroscopy
FOF	Fear of falling

FOV	Field of view
FPS	First-person shooting
FREE	Fiber reinforced elastomeric enclosure
GPS	Global positioning system
GUI	Graphical user interface
HCI	Human-computer interaction
HMD	Head-mounted display
HOA	Healthy old adult
HRI	Human-robot interface
HRV	Heart rate variability
HYA	Healthy young adult
IBM	Inclusion-body myositis
ICA	Independent component analysis
ICs	Independent components
IMU	Inertial measurement unit
ITQ	Immersive tendencies questionnaire
KKse	Kitchen knife safety educator
LWR	Lightweight robot
MCI	Mild cognitive impairments
MER	Music emotion recognition
MIX	Mason Innovation Exchange at George Mason University in Fairfax
ML	Machine learning
MMORPG	Massively multiplayer online role-playing game
MT	Movement time
MU	Movement unit
NUI	Natural user interface
OST	Optical see through
PATH	Path length
PAV	Peak angular velocity
PbD	Programming by demonstration
PQ	Presence questionnaire
PSD	Power spectrum density
QUEAD	Questionnaire for the evaluation of physical assistive devices
RAS	Rhythmic auditory stimulation
ROI	Region of interest
ROM	Range of motion
ROS	Robot operating system

SDT	Self-determination theory
SME	Mmall and medium-sized enterprise
smQ	Self-report questionnaire
STC	Sinusoidal transform coding
STV	Spatio-temporal variability
TLX	NASA task load index
TUI	Tangible user interface
UCD	University city district
UI	User interface
VAS	Visual analog scale
VLE	Virtual learning environment
VR	Virtual reality
VR4VR	Virtual Reality for Vocational Rehabilitation System at the Center for Assistive, Rehabilitation and Robotics Technologies, University of South Florida
VR-CET	Virtual reality-based cue exposure therapy
VST	Video see trough
vYv	VoiceYourView
WAM	Whole arm manipulator
WMSD	Work-related musculoskeletal disorder
XR	Extended reality
XRG Lab	Extended Reality and Games Laboratory at the School of Information, The University of Arizona

1

Recent Application Areas, Interaction Techniques and User Interfaces in Virtual Reality

Yafei Xu, Limin Zhang and Lila Bozgeyikli*

School of Information, University of Arizona, USA
E-mail: lboz@email.arizona.edu
*Corresponding Author

1.1 Introduction

Virtual reality (VR) has been actively used in a wide array of areas in recent years, thanks to the advancements in hardware which enabled affordable headsets with high technical capabilities (e.g., 6 degrees of freedom position tracking, standalone systems). This chapter aims to give a snapshot of the recent VR studies with an emphasis on novel system components or application areas. Mainly, the use of the following topics in VR will be discussed: education and training, assessment and treatment of cognitive disorders and eating disorders, increasing empathy, entertainment, interaction and user interfaces (UIs).

1.2 Education and Training with Virtual Reality

1.2.1 Language Learning in VR

Although people intend to learn languages for many purposes, there is still much less of them who can sustain the learning path. The current language learning tools cannot bring the fun part of immersive learning experience while studying abroad is not a privilege to everyone. The truth is, it is important to acquire a language in the context within its related culture. In recent years, the affordable VR tools for immersive experience have arisen in the application areas of education and tourism. Cheng et al. explored teaching language and culture with a VR game [1]. Their work focused on their design process of a VR game adapted from "Crystallize" by Culbertson et al. [2], which is a 3D video game for learning the Japanese language. They built the game in VR with Oculus Rift. Particularly, they investigated the use of VR in designing the game mechanics for physical cultural interaction like bowing in Japanese greetings. The authors performed a formative user study with 68 participants to evaluate the design. Their primary evidence was that there was a statistically significant increasing sense of Japanese cultural involvement in the players after they were trained with the VR games on how and when to bow. For the language learning outcomes, there was no evidence for statistically significant difference. The user evaluation showed positive effects of VR even though the VR headset itself caused some cybersickness in players which had no proved effect on the learning process. In their paper, it has been demonstrated that the design of language learning and VR games can to be improved by integrating physical culture learning, such as bowing. The existing complex confounding variables limited the precise conclusion of the general effects of VR on this topic. Instead, the authors analyzed the experiment to assess the design of the game with VR. They also mentioned that the enthusiasm of the players for VR might have affected their attitude towards the designed game and may count as a bias. Nevertheless, the authors concluded that VR can be beneficial for this scenario and provide motivation for language learners. Future work can be identified by increasing the sense of presence, such as primarily using speech recognition as an input mechanism. The authors also planned to test both VR and non-VR game versions on players to see which version could be preferred. With respect to literature, further exploration can be done using 3D reconstruction technology [3] to improve engagement by adding non-player-controlled characters and interactive scenes to gamify the featured locations for learning experiences.

Embodied learning plays an important role in language acquisition. Previous studies showed no difference in body expressions between first and second language [4], although learning the second language was found to be mainly affected by audiovisual elements [5, 6]. Due to capabilities of stand-alone tracking, real-time feedback, and immersive virtual environments, VR has become a preferred tool for kinesthetic language learning. Kinesthetic language learning refers to learning language through bodily activities connecting the mind and the body. With these motives, Vázquez et al. used VR for kinesthetic language learning with a system named Words in Motion, which could capture user actions and offer real-time feedback [7]. The feedback of the study was the word in the aimed second language according to its corresponding action implication. Therefore, the Words in Motion VR system was used for second language vocabulary learning kinesthetically. A user study with 57 participants was conducted to assess the usefulness of the developed VR kinesthetic learning system. Participants were randomly divided into kinesthetic VR, non-kinesthetic VR, and text-only groups. A set of 20 challenging Spanish transitive verbs were selected for the experiment. Participants were then evaluated by tests that asked them to translate the words in English immediately after the training and one week later. The results indicated that the participants in the text-only group showed better performance at the beginning, which could be explained by the highly immersive and high-fidelity features of VR as a tool in learning in alignment with the previous work. Both VR groups showed no difference immediately after the training. After one week, participants in the VR kinesthetic group had similar performance as the participants in the text-only condition. They showed significantly better performance than the participants in the non-kinesthetic VR group. The retention rate was significantly increased by the included kinesthetic elements in VR. The authors also found a positive correlation between the word-action pair execution time and word remembering time, meaning that VR could positively enhance the language learning process by providing the kinesthetic aspect. In the end, the authors concluded that kinesthetic language learning in VR could be an effective language education tool which adds more exposure.

1.2.2 Learning Physics Concepts in VR

Physics is among the most challenging and important courses for students [8]. Grivokostopoulou et al. explored creating a physics education in VR for students and teachers of K-12 level [9]. The authors wanted to explore the innovative instruction approach using 3D nature of VR to better attract and assist students in deep learning and understanding physics concepts through virtual experiments, laboratory procedures and physical processes. Pedagogical virtual agents were also created to lead and assist students by analyzing and explaining the scientific processes in the virtual environment. Spatial cognition is known to be enhanced with VR [10]. Simulations can induce better understanding and engagement of students with learning difficultly to grasp concepts [11]. With these motivations, the authors thought that virtual laboratories could save expense of purchasing physical instruments and provide students with remote learning possibilities without time constraints. Virtual agents gave the students a higher sense of presence while interacting with the virtual world. Analysis on physics topics was conducted in two stages. The first stage was to specify the content and the second stage was to specify the difficulty of each topic by analysis. In total, 33 subtopics were used. Three experienced physics teachers participated in the study to evaluate the challenges of each topic. The most challenging topic for students was identified to be electricity and magnetism. The least challenging topic was found to be the structure of matter. The authors suggested their future work as gamification in learning, and utilization of feedback sequencing framework, large scale evaluation, assessment of multiple learning situations, and data mining to better analyze behavior of students with learning activities in VR environments.

1.2.3 Learning How to use a Cooking Knife in VR Through Intuitive Tangible User Interfaces

VR has been an excellent medium for training with high safety, since it reduces the risks and consequences of mistakes that are made in the virtual environments compared to real-world places. A recent work that explored safety and communication issues when teaching children how to use a cooking knife was carried out by Saito et al. [12]. The authors developed a VR-based education system that included tangible user interfaces (TUIs) to see if VR would provide a better training alternative due to increased safety. The system was named "Kitchen Knife Safety Educator (KKse)" and was composed of a knife device, a cutting board device, and an ingredient device. The KKse was a portable device with a haptic force feedback system with virtual knife and food, aiming to teach children how to use the "thrusting cut" method with a knife and how to make use of the ingredients. The use of each device component of the KKse system was described as follows. The "thrusting cut" was a basic Japanese cutting method, which included cutting forward with a thrusting motion. This cutting method used less force and provided more safety in the real world. The knife device in the system gave haptic feedback from the regulated tension based on the hardness of the cut ingredient. The ingredient device was made up of four pressure sensors to receive the force information from the virtual knife, and a conductive sheet to detect if the knife touched the user's fingers. A notification sound would be triggered when the cutting task was successfully finished. The cutting board device was used to check the stable positioning of the food items to instill safe cutting practices. The authors pointed out future research directions as performing user studies to validate the effectiveness of the proposed KKse system. This recent work demonstrates the use of highly immersive VR in conjunction with intuitive TUIs to provide safe training with high transfer rates to the real-world in daily life activities, such as cooking.

1.3 Assessment and Treatment of Cognitive Disorders in VR

1.3.1 A Review on Recent Works on the Assessment and Treatment of Cognitive Disorders in VR

Cognitive health is closely related to an individual's environment, which opens up the possibility of novel applications in VR in this area that exploits the highly engaging and immersive nature of VR. Recently, in their survey paper on recent works on cognitive health in VR, Freeman et al. conducted a systematic review of pioneering empirical studies on VR in mental health [13]. In the paper, the potential application areas of VR in mental health disorders were described and discussed. The literature review was carried out using PubMed to search peer-reviewed publications before 2017. The researchers had come up with their own inclusion and exclusion criteria for the selection of the identified papers. Overall, they identified 285 studies including 86 concerning assessment, 45 theory development, and 154 treatment. About 192 studies of anxiety, 44 studies of schizophrenia, 22 studies of substance-related disorders, and 18 studies of eating disorders were mainly researched. They found out that the studies of VR in mental health disorders were in an early stage and lacked methodological quality. There still existed a big gap between useful applications of VR in mental health. As an effective application area, they found out that anxiety disorders could be reduced by VR exposure treatment with plenty of methods. Also, the authors found out that VR had been misused as a term in many fields for experiences that included no immersive or interactive techniques, which was the reason why many studies were excluded from their literature review. After their review, the authors identified the high level of fidelity and high capability of engaging content delivery in VR as key factors that opened up the possibilities of psychological treatment in this medium. The authors pointed out that the recent interest in VR fed interest in neuroimaging as well. In the end, the authors suggested that VR was a potentially effective medium for the understanding, treatment, and assessment of cognitive disorders although being at a relatively early stage. The studies that the authors had reviewed along with their discussions are included in the following subsections.

Most VR studies focused on treating anxiety disorders including 127 intervention reports, 46 studies on treatment validation in VR, and 19 studies that invested the casualty of anxiety. Those studies mainly focused on specific phobias, social anxiety or post-traumatic stress disorder. Exposure intervention in VR with an accompanying therapist was the most used method. The authors of those studies mainly used case studies and randomized controlled experiments, which was found to be poorly designed, in Freeman's literature review. VR treatment seemed to have an equal effect on treatment as compared to face-to-face intervention. The treatment effect had been evidenced to be significantly important and had a high transfer rate to the real world. The authors pointed out that the drop-out rates were lower when using the VR method. The specific types of VR techniques that were used in the mentioned studies were not clearly mentioned. The authors pointed out to increased sense of presence and immersive audio in VR as the aspects that induced increased anxiety in users.

Depression: Only two studies that the authors had included in their review used VR for the treatment of depression. In these studies, VR was shown to be effective in decreasing the level of depression in participants.

Psychosis: There were several studies about psychosis in VR: 23 studies of theory development, 15 studies of assessment, and 6 studies of testing. Psychosis is known to be a complex problem; hence, the use of VR in this area yielded varying results. In this area, VR was primarily used to understand the causality by assessing the disorder experiences, which was known to be a safe approach. VR was suitable for conducting paranoia assessment since neutral social circumstances made it easier for the users to detect hostility. In an example, which demonstrates the possibility of implementation of concepts in VR that could not be implemented in real life, the researchers had created a VR scenario in which they manipulated the height of people to adjust their self-esteem and cause paranoia. VR was found to be effective in manipulating the factors that affected paranoia. As a drawback, the studies in this sub-domain had small sample sizes and data. In the literature review, mania was found to be unexplored in VR.

Substance Disorders: VR was used in the literature to simulate cues that led to substance use disorders, such as alcohol abuse, drug misuse or excessive gambling. In total, 22 studies were found: 15 in assessment, 5 in treatment and 2 in theory development. A lot of previous studies had shown that cravings could be triggered successfully in VR. Misuse of different kinds of substances was studied vastly in the literature. Most related works were performed on smoking, where the VR environment could lead to significant cravings for cigarettes. Previous studies reported that VR was promising in decreasing smoking cravings. Randomized controlled experiments for smoking cessation are currently undergoing to provide scientifically powerful conclusions.

Eating Disorders: VR was reported to be effective for the treatment of eating disorders, such as decreased cravings for unnecessary food intake, altered body image, increased emotions, etc. In their review, the authors found 18 studies in this sub-domain: 10 performed on treatment, 7 on assessment and 1 on theory development. Early studies in this area lacked scientifically sound methodology. In the studies, cravings for food in VR were comparable to real food cravings. Augmented reality has also been an area of interest for the treatment of eating disorders. A recent study used augmented reality to present high-calorie food to individuals with under-eating disorders. The data from this experiment demonstrated that body image could be improved by VR along with standard cognitive behavioral therapy (CBT). The patients with anorexia nervosa were exposed to inclusion-body myositis (IBM) in VR, which led them to have normal body estimation for 2 hours. To sum up, VR was found to be a promising medium, for the treatment of eating disorders, mainly due to its highly immersive nature.

Other Disorders: Potential areas for using VR were pointed out in the previous studies, such as therapies of sexual disorders (e.g., desire and arousal) and treatment of sleeping disorders. Although no studies were found in these areas, a study that used a VR paradigm to assess the bad effects of sleeping disorders in children was conducted. These areas constituted promising future work domains that were identified in the previous studies by the researchers.

1.3.2 Improving Lives of Individuals with Alzheimer's Disease with VR

In an effort to improve the lives of individuals with Alzheimer's disease, Giovannetti et al. used VR to collect the finger movement data of patients who have everyday action impairments (EAI) [14]. The authors pointed out the promising nature of VR in data visualization for these purposes. Building upon, other researchers have focused on data mining and therapy of mild cognitive impairments (MCI) in VR [15]. Bringing this one step further, Martono et al. wanted to know if the clusters formed by the time series data mining algorithm would work well with the finger movement data obtained from VR applications [14]. If so, they could use such an analysis to improve the treatment of patients with EAI by clustering those patients using their symptoms in the future. In their study, the researchers recorded finger position data from 10 typical participants while they performed everyday tasks in a virtual kitchen, and then transformed this data into finger acceleration data. Time series data mining algorithms were used on these data to cluster the participants and performance-based assessment was performed to validate the clusters qualitatively. As a result, the clusters were well-formed and displayed a nice cluster dendrogram even though the authors acknowledged that finer partitions could have been created to achieve more refined clusters. The authors also characterized the clusters with prediction density estimates. In terms of the prediction densities, the dataset for cluster one showed a small variability while the dataset for cluster three had a high variability. Their correlated performance-based measurement showed the length of performance time, the occurred types and amount of errors made by the participants in the performed everyday tasks. It turned out that the participants of each cluster had similar performance time, error types, and numbers. From these results, the authors concluded that their cluster methods using time series data mining algorithm could work well with motion data that is collected from VR-based applications even though they acknowledged that more separation and participants with real EAI symptoms should be involved in order to make scientifically stronger conclusions. Possible further improvements to the study were pointed out as using hand speed or movement data.

1.4 Assessment and Treatment of Eating Disorders with VR

In recent years, VR has been used as an aiding tool in the assessment and treatment of over-eating disorders [16–19] and under-eating disorders [20–23]. This section covers recent noteworthy studies on VR's use in the assessment and treatment of eating disorders.

1.4.1 Cue Exposure Therapy in VR for Undereating Disorders

Several previous studies focused on using VR for exposure therapy since it is a practical way of inducing a wide array of virtual stimuli. In their recent work, Pla-Sanjuanelo et al. evaluated VR-based software that assisted people with bulimia nervosa (BN) and binge eating disorder (BED) with cue exposure therapy (CET) [24]. The authors assessed the software's capability in eliciting two responses: food craving and anxiety towards food photos.

They also probed into which one of the two responses differentiated more between the two groups. Approximately, 58 outpatients with BN and BED and 135 healthy undergraduate students were recruited as the clinical group and the control group, respectively. The frequency and severity of binge episodes were assessed with the Eating disorder inventory -3 (EDI-3) [25]. This study focused on the effectiveness of non-immersive virtual environments in a quiet darkened room. All participants rated 30 images in 4 environments (i.e., kitchen, dining room, bedroom, and cafeteria) in terms of the induced level of craving for food. Each participant's top 10 rated images were selected for their own stimuli, and mixed with a neutral one (i.e., a virtual stapler) randomly. Those 11 images were present in each of the four environments, thus a total of 44 images were displayed for viewing and assessment. Once the participants viewed one image for 20 seconds, they scaled the level of food craving and anxiety on two independent visual analog scales (VAS) from 0 to 100. The results showed that both groups had higher levels of food craving and anxiety than being exposed to the neutral cue. The level of food craving and anxiety were higher in outpatients as compared to healthy individuals. Cue-elicited anxiety was found to be better than cue-elicited craving in differentiating the two groups (i.e., clinical and healthy individuals). The study demonstrated the effectiveness of VR in the treatment of eating disorders (i.e., binge eating and bulimia). The level of craving and anxiety were suggested as a design consideration for future CET programs. The authors had pointed out future research directions as exploring more immersive virtual environments for the same purpose.

Another notable recent example study was performed by Pla-Sanjuanelo et al. where they investigated the efficacy of two second-line treatments in patients with BN and BED who had resistance to CBT: VR-CET alone, and the combination pf VR-CET and pharmacotherapy [26]. Previous medical studies had shown that cognitive-behavioral therapy (CBT), the first-choice treatment for BN and BED had no effect on a considerable portion of the patients. Antidepressant medication has been recommended as an alternative to or in combination with CBT for these patients. Cue-exposure therapy (CET) has been considered as another choice with potentially fewer side effects. The VR's value in CET lies in its cost-effectiveness and flexibility in custom-based scenario creation and alteration. Several previous works have demonstrated VR-CET's compatibility with CET in several important aspects. In this recent study, the authors tried to explore a deeper understanding of VR-CET in an experiment that compared VR-CET and VR-CET combined with antidepressant medicine. 32 BN or BED patients who still experienced resistance to CBT were referred by hospitals for participation in the study. 17 of them took a constant dose of antidepressant medication at the beginning of each VR-CET session, while 15 individuals did not. Both groups' interventions had six individual sessions which tool 60-minutes and held twice-weekly over three weeks. In the first session, the top ten combination scenarios (environment and food) that cued a participant's highest food craving were selected and used for that specific participant for the rest of the experiment. During the sessions, craving and anxiety levels were rated by the participants periodically. When the reported anxiety decreased by 40%, the session ended. Core behavioral features were collected for assessment during the two weeks prior to the beginning of the second-line intervention, during the two weeks after the end of the intervention, and during the two weeks after the 6-month follow-up. EDI-3 and FCQ-T/S (the State and Trait Food Craving Questionnaire) were employed at the end of the intervention and at the follow-up. Results showed a statistically significant difference in the intervention outcomes tested at the 3-time points in both groups, but no significant difference between the two groups. Hence, the addition of antidepressant medicine into VR-CET did not provide any additional benefit in patients. The study contributed to the existing literature in VR's effectiveness in the treatment of eating disorders from a different perspective, which can save the patients from various side effects that are associated with the antidepressant medications in cases where these medications do not contribute to the treatment.

1.4.2 Using VR for Jogging for Exposure to Acute Urge to be Physically Active in Patients with Eating Disorders

VR has been effectively used in the treatment of eating disorders, mostly with a focus on cue-exposure therapy and treatment for body image disturbance. Recently, researchers have started to explore handling implications of eating disorders with VR. A study in this area was conducted by Paslakis et al. where they explored the use of VR as a therapeutic tool for patients with eating disorders who suffer from the acute urge to be physically active [27]. This phenomenon of engagement in excessive physical activity was reported by the majority of patients across all types of eating disorders [28]. Hence, Paslakis et al. designed a pilot study as a proof of concept to overcome this. Twenty female patients (10 with anorexia nervosa and 10 with bulimia nervosa) participated in this study. All of them wore VR goggles, watched a jogging video and tried to engage in simple movements.

During the sessions, the acute state was assessed at 9-time points (0, 4, 8, 12, 16, 20, 24, 28, and 32 minutes) with a 10-item self-report questionnaire (smQ) assessing the cognitive, emotional, and behavioral aspects. Participants had the choice to end the session at the 8th, 16th and 24th minutes. The assessments of the acute urge were collected at the beginning and the end of the sessions. Besides, other variables, including saliva samples, blood samples and BMI were collected before and after the exposure. Overall, there was significant reduction in the scores of smQ from baseline to postexposure, which provided preliminary evidence supporting the feasibility and benefits of applying VR as a treatment for the acute urge for engaging in physical activity that is seen in eating disorders. The authors pointed out the next step as an experiment with a larger sample size to verify the findings.

1.4.3 Outcome at Six Month Follow Up After VR Therapy for Undereating Disorders

It has been an ongoing debate whether VR therapy would be sustainable in the long term or not. As it has been some time now since more accessible VR systems have been more widely used in different areas, researchers have started to access follow-up data after VR-based therapies. Such a research was recently conducted by Ferrer-Garcia et al. where they compared two second-level treatments for BN and BED: VR-based cue exposure therapy (VR-CET) versus additional cognitive behavioral therapy (A-CBT) [29]. Previously in 2016, a randomized controlled study was conducted by the researchers at five clinical sites in three Europen cities [30]. The results in that study had showed improvements in all patients treated either by VR-CET or A-CBT. Overall, patients in the VR-CET group had showed better short-term results, especially, a substantial decrease in the frequency of binge and purge episodes and self-reported tendency to engage in episodes of overeating, food craving, and anxiety. In addition, the rates of abstinence from binging and purging had been also significantly higher in the VR-CET group. The recent paper investigated whether the outstanding performance of VR-CET could be maintained in the long-term. Fifty-eight out of 64 patients from the previous study completed the 6-month follow-up assessment. The results indicated both treatment effects were maintained at a 6-month follow-up. Similar to the short-term outcomes, VR-CET showed more promising results in the long-term, especially for one index: the binge-purging abstinence rate. The authors pointed out the importance of understanding the underlying conceptual model and mechanisms in VR-CET, which calls for further research that will enable and encouraged researchers for future clinical applications.

Most of the studies thus far have used VR for the assessment and treatment of eating disorders targeted patients. It has been recently reported that 117 million American adults (around 50% of the whole American adult population) have one or more preventable and chronic diseases that is related to poor quality eating in terms of nutritional variety or number of calories, and lack of physical activity [31]. Until now the baseline in the area of VR and eating disorders is mostly established, the possible future work in this area can include preventive studies, such as instilling healthy eating habits to individuals to decrease their predisposal to preventable diseases.

1.5 Use of VR for Increasing Empathy and Perspective Taking Ability

VR has been successfully used as a tool for perspective-taking, helping individuals to put themselves in other's shoes or in the shoes of their future selves through controlling or interacting with virtual avatars that looked very similar to the individuals' real selves. These studies were in different contexts, such as reducing ageism by exposing individuals to self-images that were made to look aged by image processing [32], reducing public speaking anxiety by showing individuals their virtual replicas that gave a public speech successfully [33], improving the sense of empathy towards challenged populations and increasing their willingness to help others [34], promoting exercise by showing a virtual avatar that gained weight if the user stayed inactive and lost weight if the user exercised [35], showing virtual representations of individuals as exercising [36], and positively influencing future financial decision making through promoting savings for the future by showing individuals aged virtual avatars that looked like them [37].

A sense of presence is an important user experience aspect in VR, which can be described as the sense of being in a virtual world and forgetting the real-world surroundings. It has been found that individuals identified themselves more with virtual avatars that looked similar to them as compared to dissimilar virtual avatars, and it created an increased sense of presence in individuals [38–41]. Bandura's social cognitive theory states that greater similarity and identification with a model causes increased social learning and imitation of the modeled behaviors [42]. Hence, many of the recent studies employed virtual avatars that looked similar to the individuals who used the system. These studies that were successful in improving behavior through highly embodied experiences through the use of virtual characters that looked similar to the individuals in appearance make VR a promising tool in enhancing individuals' knowledge on important topics, such as health, well-being and making conscious decisions regarding the future. Previous research shows that many individuals are characterized to fail to identify with their future selves, possibly because of the lack of imagination or false belief [43, 44]. Moreover, it has been suggested that individuals may experience empathy gaps and may not judge correctly how they will feel about the effects of their current decisions on their future selves [45]. It has been suggested that more vivid the images of a future event, the more intense the emotions associated with thinking about the outcome and the possibility of that outcome would be in individuals [46]. As it can be observed through the mentioned previous studies, VR is a powerful tool in creating a connection with individuals' future selves and creating awareness in terms of seriousness and likeliness of the implications of current choices on their future selves, which is difficult to achieve through conventional methods in real life.

Due to these powerful embodied experiences it provides, VR was proven to be more effective in creating improvements in individuals' habits compared with traditional methods such as video and print-out information channels. In a recent study, Ahn et al. found out that individuals who cut a virtual tree in an immersive virtual environment were observed to consume 20% less paper than participants who read a print description of the tree cutting process [47].

1.5.1 Preventing Bullying with VR

In most bullying scenarios, intervention by bystanders was found to be a very effective resolution to stop peer bullying [48]. VR was found to be a promising medium to provide such an immersiveness and increase in the feelings of empathy. In their recent research, McEvoy et al. conducted two studies comparing the effects of interventions in bystander-focused bullying situations using customized VR, non-customized VR, and video [49]. In the experiments, 78 participants were randomly distributed to one of the mentioned conditions. Three conditions used the same campaign video. In the video environment, the participants could not move or interact with the environment. In the customized VR environment, the participants saw the victim wearing a shirt with the same university logo and color as the participant wore. However, the victim in the non-customized VR environment wore a shirt in a different color without a university logo. The researchers measured the level of empathy, attitudes on being a victim and bullying, perceptions, future actions, and presence. The measures on empathy and perception of bullying received the highest scores in the video environment with statistically significant differences. The authors also performed a follow-up focus group study with ten participants to investigate the above results and to compile suggestions on building better VR simulations for this purpose. The factors to induce higher empathy in VR were identified as photorealistic graphics, interactive functions, and tailored customization.

1.6 Use of VR for Entertainment-Based Activities

1.6.1 VR as an In-Car Entertainment Medium

VR is known to cause cybersickness when used inside moving vehicles due to the virtual movement that is based on the sensors of the head-mounted displays (HMDs) and the real movement that is based on the physical world as well as information on the simulated visual and vestibular systems are not congruent. Nowadays, long commute times are very common for everyday working people, therefore, gaming in mobile VR gets the potential with a large consumer market to help commuters to spend idle time in an enjoyable way. With these motivations, Hock et al. developed a prototype that enabled more comfortable use of mobile VR in moving vehicles through the subtraction of the car's rotation and the mapping of the vehicular movements with visual information [50]. This provided users a sense of more accurate kinesthetic forces in the VR environment. The authors designed the vehicle's movements and the visual information in VR to eliminate cybersickness as much as they could, based on previously recommended practices in the literature. A user study with 21 participants was performed to compare the user experience in CarVR in moving and static environments. It turned out that users felt significantly increased enjoyment and immersion with reduced simulator sickness when using the CarVR in moving environments, as compared to the baseline scenario. The VR condition was found to be more fun by the users. Simulator sickness, engagement, enjoyment, and immersion were extracted by SSQ, E^2 I and a comparative questionnaire. The SSQ score was found to be inaccurate by the authors, mainly due to the fact that the SSQ was designed for severe movement conditions (e.g., flight simulators). Hence, the difference in the SSQ measures was insignificant between the two versions.

Most of the participants found the VR condition to be more exciting and entertaining compared with the control condition, which was consistent with the previous studies which showed that perceptual action that was consistent with visual information increased immersion [51]. Braking was the most uncomfortable situation reported from the user questionnaire, which was expected since braking was not an anticipated action that would increase simulator sickness in users. The study only used a single type of track. Future research areas were identified as multiple types of tracks, large sample size, more diverse major distribution in participants (in the mentioned study, 10 participants out of 21 were students majoring in computer science), exploring the effects of sear position on the user experience, level design, different types of force shifts, and exploring a more granular way of detecting simulator sickness than the SSQ. The authors also investigated the design space of VR applications with real kinesthetic forces for entertainment in moving vehicles and gave their design considerations to help future developers. The study explored the use of VR as an entertainment tool in moving vehicles by reducing cybersickness, which constitutes the most significant barrier in VR's use in such environments. With similar improvements, in the near future, mobile VR can be the main entertainment tool for commuters to spend time in an enjoyable way while traveling.

1.6.2 Shopping in VR

Consumer experiences have been transforming into more virtual forms in recent years. Examples can be given as real estate tours, virtual home construction demonstrations and virtual try-on of various items, such as makeup and clothing. In their work, Kerrebroeck et al. explored the use of VR in consumer experiences in a shopping mall, mainly focusing on perceived crowdedness [52]. In a previous study, smart technologies were demonstrated to improve shopping experience [53]. This paper introduced the research of retail atmospherics in VR. It also extended the application of VR in marketing and retailing by Renko et al. [54]. Escapism can be described as the tendency to leave the real world and related problems [55]. Building upon this, the researchers utilized an immersive VR environment to provide the customers a sense of escapism from the crowd. In this study, a 2×2 quasi-experimental between-subjects design was used. The control group had regular shoppers and the experimental group had shoppers in VR. Participants were randomly picked, and a questionnaire was completed following the experiment. Low or high levels of crowdedness were induced in the study. A Christmas-themed VR experience and related settings were created in a shopping mall. The Oculus Rift DK2 was used as the VR headset. The regular shoppers saw the regular Christmas decoration without any VR experience. There were 103 VR users and 80 regular shoppers who participated in this study. To test the hypotheses, the researchers performed two-way analyses of variance (ANOVA). "Attitude toward the mall", "approach behavior", "mall satisfaction" and "loyalty intentions" were used as the dependent variables. As a result, participants who were involved in the VR experience showed more positive reactions. The perceived crowdedness level was higher, the impact on the mall attitudes, satisfaction and loyalty were greater. Some limitations of this study were worth to note. Since self-selection bias may exist in the VR users for data collection, consumers who tend to stay at home to avoid the crowds at shopping malls should also be included. The exposure duration of the VR experience in the study was short, and the results were dependent on the task design. The researchers concluded that relaxing and pleasant VR applications could be valuable in crowded business places for improved user experience, similar to shopping malls.

1.6.3 VR for Parasailing

It is known that physical stimuli in VR can improve the sense of self-motion. However, to date, motion simulator-focused studies were limited to the specific workspaces and mostly focused on horizontal motion. To address this gap, Kang et al. studied the stimulation of floating sensation through vertical movement [56]. The authors created a virtual parasailing system where a VR headset provided visual stimuli and cable wires provided physical stimuli. A washout filter was used to depict physical acceleration. Vertical acceleration and deceleration were included in the virtual parasailing environment to investigate the natural integration of visual and physical stimuli. The results were then tested through various full-course virtual parasailing scenarios. General comments were collected from subjects through interviews. It turned out that a large range of differences in visual and physical stimuli could be accepted by the users and the users' attitudes were crucial in deciding the amount of gains. The researchers pointed out that the research was applicable to other floatation simulations, such as jumping and free-fall motions. Future work was pointed out as investigating the applications of free-flying extreme sports or games in VR using the developed acceleration/deceleration controls. Haptic cues could be integrated into the system in future versions to increase the amounts of feedback and visual gains. The study served as a demonstration of an interesting current application area of VR (i.e., parasailing).

1.7 Interaction and User Interfaces in VR

1.7.1 Touch-Based UI for Mobile VR

Recently, VR has become much more accessible and affordable through the emergence of consumer-focused systems by companies such as Oculus and HTC. There is a great potential for mobile VR to be more prevalently used in smartphones, increasing the possibilities of experiences for various purposes. VR is described to be one of the new communication tools in the current age, which requires new interfaces and meta mediums to be seamlessly integrated into the existing technological devices [57]. With these motivations, Lee et al. investigated a new touch-based user interface (UI) device in the form of a touchpad that was mounted in front of a VR headset [58]. In their study, the users could intuitively locate and manipulate virtual targets by seeing through the headset and touching on the front pad precisely and easily using their sense of proprioception. Other advantages of the proposed system were identified as no significant monetary, weight and power cost that was incurred by the inclusion of the touchpad interface on a VR headset, while the addition of the touch sensing technique providing benefits in user experience in terms of interaction. After the preliminary investigation on the users' intuition and accuracy while using the front touchpad, different VR UI design options, including a binary selection, a typical menu layout, and a keyboard, were compared. New front-touch interactions, such as Two-Fingers for quick selection and Drag-n-Tap for accurate selection, were investigated in this study. The authors enlarged the interaction space to allow for more types of interaction techniques using the front touch area. User studies were performed using the VR keyboard for text-input and menu selection to assess the usability of the proposed front touch interface. The results demonstrated improved intuition and performance as well as decreased cybersickness. The participants stated a preference for the proposed touch interface. The front touch interaction was proven to be an effective and intuitive extension for the mobile VR headset. The researchers pointed out future research directions as improving the interface and discovering more design possibilities for improved user experience.

1.7.2 Mobile Ungrounded Force Haptic Feedback in VR

Haptic feedback in interaction has been an active research area in VR for a long time. The majority of the previous research was focused on utilizing grounded force haptic feedback in stationary environments.

Recently, researchers have started to explore unique applications of haptic feedback for improved VR interaction. A notable example is Heo et al. recent previous work, where they explored a mobile ungrounded force haptic feedback device that was capable of applying high-magnitude force in arbitrary directions [59]. To demonstrate this, the authors had created the Thor's Hammer, which used propellers to give ungrounded forces in 3 degrees-of-freedom (DOF). The researchers performed a technical evaluation of the Thor's Hammer, showing that up to 4 N of force could be applied on the hammer in arbitrary directions with no more than 0.11 N and 3.9-degrees of average magnitude and orientation errors. By assessing the force control accuracy, responsiveness, noise level, and power consumption, the researchers evaluated and discussed the design suggestions regarding the prototype. Based on the advantages of the mobile Thor's Hammer that could create strong and continuous force-based haptic feedback with high precision, the researchers had created four VR scenarios to incorporate the use of Thor's Hammer. In the "Feeling the water" application, the users used Thor's Hammer to feel the water flow with varying strength and magnitude. The "Herding a sheep" application made the users use the Thor's Hammer to feel continuous and precise forces in varying magnitudes and directions while they were pulling a moving lamb. The "Pushing buttons" application made the users feel various rigidities of buttons within the scene with the Thor's Hammer. The "Simulating different weights" application allows the users to feel different dynamic forces with various gravity values. The qualitative user study was then conducted to understand effects of the applications of the Thor's Hammer in VR, which induced haptic force feedback, on user experience. 6 participants were recruited and tested the four VR applications with or without using the Thor's Hammer for haptic feedback. After each test, the participants filled out a questionnaire and, in the end, they picked their most liked VR application and justified their choices. Overall, the participants reported that the force feedback given by Thor's Hammer made their experience more immersive and fun. The simulated forces were perceived to be strong and continuous in various directions, and to have good precision. However, the participants pointed out that the device's fan was noisy and distracting, which reduced their sense of presence. In the future, the researchers decided to address the shortcomings of the device, such as decreasing the size and weight to avoid user fatigue, as well as decreasing high latency in feedback, exploring a 6-DOF force feedback device, comparing the effects of propeller propulsion with a grounded force feedback device, and investigating the influence of unintended haptic force feedback.

1.7.3 Including Virtual Representation of Hands while Typing in VR with a Physical Keyboard

In recent years, VR has been widely used in entertainment and industry, thanks to the advancements in the hardware and numerous scientific studies that investigated novel systems or uses of VR along with user experience [60, 61]. One caveat in the mass adoption of VR was identified as the obstructed user view. The occlusive nature of VR headsets made users frustrated to use a physical keyboard and mouse, which are common fundamental interaction devices in human-computer interface [62]. To explore this, Knierim et al. developed a physical keyboard in VR to help users benefit from typing with seeing the virtual representation of their hands in the virtual world [63]. The authors developed an apparatus to track the keyboard and the fingers of users in real-time to provide a matching visual representation in VR. A comparative user study evaluated the effects of varying degrees of transparency of the virtual hand representations on the performance of typing. Sixteen experienced typists and 16 inexperienced typists were participated in the text input evaluation study with two conditions. The participants typed inside the VR, seeing the virtual representation of their hands with different transparencies or typed outside of VR. Typing experiences were explored with the assessments of task load and sense of presence. Hand and transparency variables were nested within-subject. The typing experience was the between-subjects variable. A mixed nested factorial experimental design was conducted. The text input task consisted of three sets of 10 phrases, to let the participants input these phrases as fast and accurately as they were asked. The NASA-TLX [64] and the presence questionnaire (PQ) [65] were then filled out by the participants after each task. In the end, the participants were asked to leave their comments on the typing experience. For the experienced typists, the avatar hands made no difference in their typing speed. The inexperienced typists typed only 5.6 words-per-minute slower when using semi-transparent hands. It was considered to be crucial, especially for inexperienced typists, to see optimized avatar hands in VR to yield better typing performance. Future work was identified by the researchers as exploring further optimization of the proposed method for improved user performance. The study demonstrated the importance of interaction design in VR in performance and user experience.

1.7.4 Haptic Revolver for Haptic Feedback in VR

Whitmire et al. created a Haptic Revolver, which was a reconfigurable hand-held haptic controller in VR that utilized an actuated wheel under the fingertip raising and lowering to render touch contact with a virtual surface [66]. The controller also spun to render the shear forces and motion when fingertip sliding was performed on the virtual surface. The wheel could be customized using different physical textures, shapes, edges, and active elements to offer different sensations on users. The device was spatially tracked to connect the haptic elements with the virtual world accurately. When the users were playing in VR, the rendering engine gave them a corresponding haptic response under their fingers. Two perceptual user studies with 12 participants were carried out to evaluate the effects of wheel speed and direction on the perceived realism, which later gave suggestions on the design of the Haptic Revolver rendering. In the first and the second studies, the users slid their fingers on a surface horizontally and traced paths. In the third study, 11 participants were recruited, and the user feedback was collected through semi-structured interviews to help the authors compare the Haptic Revolver with a standard vibrotactile notification in three sample applications. In the card table application, the users could touch objects of different textures in the virtual world by using the rendered wheels. In the painting and sculpting application, shapes and forces were applied on the wheels. The users could press on the wheel to use the painting tools. The third keyboard application made use of the rendered edges and shapes on the wheels to let the users feel the edges of the keys. There were significantly higher ratings for the Haptic Revolver than the Vibrotactile form, in all three applications. This implied that the Haptic Revolver had a higher-fidelity haptic rendering which would improve user experience. The authors pointed out the following future research directions: exploring automated switching between the wheels and the effects of the physical haptic response on task performance and adding a proximity sensor into the wheel for automated height calibration.

1.7.5 Bimanual Haptic Controlling in VR

A majority of the design of haptic controllers for VR in the literature was for single-hand interaction [67]. However, bimanual interactions were closer to users' needs for rendering forces. Recently, Strasnick et al. proposed a system named "Haptic Links" that served as electro-mechanical connections between VR controllers to provide different actuated stiffness [68]. Haptic Links could be rigidly attached to existing controllers with various configurations, providing specific DoF or force directions to dynamically create stiffness in a continuous range. In the Haptic Links, users perceived the forces in two hands through various forms of interaction and two-handed objects.

The researchers created and implemented three Haptic Links prototypes to offer different stiffness feedback. In the "Chain Device" prototype, a cable was pulled tight through ball-and-socket elements to make the articulated chain stiffer. The "Layer-Hinge Device" prototype controlled the rotation by locking ball joints and the distance between controllers by a hinge. The "Ratchet-Hinge Device" prototype consisted of locking ball joints and a dual-ratchet hinge, while the directions of motion could be controlled by unfastening any individual opposing pawls. The researchers conducted technical and user evaluations to analyze the pros and cons of their design and the usefulness of the proposed Haptic Links. The technical evaluations compared the specific techniques and experimented with torque-angle curves for each prototype. The user study included 12 participants to compare and rate the perceived object rendering realism in various Haptic Link devices and unlinked controllers for different object types. Four objects, rifle, bow, trombone, and pistols were explored by each user using each of the four devices. These four virtual objects were created to provide a wide range of stiffness and motion. Based on the results, the usage of different Haptic Links for various interactions and objects was suggested for the system's design. Furthermore, the researchers investigated the usefulness of the Haptic Links in providing bimanual interaction techniques in VR. Interaction for object summoning, object retrieval, object grasping, and controller grounding were explored using the Haptic Links. The Haptic Links was found out to have advantages and capabilities to provide realistic rendering of various objects by using inter-controller stiffness feedback. The Haptic Links could provide increased realism in interaction and two-handed haptic rendering in VR with various haptic tools and potential prototypes for customized user experience. The authors pointed out future research directions as improving the devices and investigating the effects of the Haptic Links on the perception of different haptic techniques and coordination between hands.

1.7.6 Data Visualization in VR

Millais et al. compared the effects of data visualization in VR with traditional 2D environments through comparative user studies [69]. As VR had started to become a potential medium for the immersive data visualization and only a few studies directly weighted the usefulness of data exploration in VR versus in traditional 2D environments, the authors performed a workload and insight-based evaluation with a user study to compare data visualization in those two circumstances. The authors created two VR data visualization modes. "Be The Data" was a three-dimensional scatter plot [70] and "Parallel Planes" was an n-dimensional dataset [71, 72]. Both VR visualizations were developed with the Unity game development engine for Android smartphones to be used with Google Daydream VR headsets and controllers. Participants could navigate and explore the VR environment from different perspectives. The equivalent 2D data visualization modes were also created for comparison. Sixteen participants were included in the study and were randomly assigned to each of the software modes (VR vs. 2D). The user study included a training stage and a main stage, which made the users familiar with the data visualizations and the "think-aloud" protocol to explore different datasets and report their insights. NASA TLX questionnaires were used to measure the perceived task workload. Audio recordings were coded to transcripts according to an existing scheme by Saraiya et al. [73]. Finally, the researchers compared the frequency of the coded items and the TLX measures between the two modes. There was no statistically significant difference in the total task-workload between the compared conditions. Nevertheless, the users in the VR condition gained more precision and depth in their insights. The results also implied that VR-based data visualization induced more satisfaction and confidence in participants. The limitation of this study was the small sample size. The future work was aimed at including more users in the study, making stronger conclusions through statistical analysis, and exploring collaborative data analysis in VR. Overall, the authors demonstrated potential usefulness of VR in data analysis.

1.7.7 Eyes-Free Object Manipulation in VR

Due to its highly visual-based nature, object manipulation in virtual and augmented reality has been linked with eye engagement. Recent studies explored leveraging people's spatial memory and sense of proprioception to enable object manipulation without the eyes being actively engaged in the experience [74, 75]. It was pointed out that the users could be more focused with less fatigue and less cybersickness in immersive VR environments with reduced head movements and interruptions [76]. Along parallel lines, Yan et al. designed an eyes-free VR system for target acquisition in VR with the aim of improved user experience [77]. The authors conducted three user studies to investigate the implications of eyes-free target acquisition in VR on user experience. In the first study, 12 participants were asked to report their acceptance of the proposed eyes-free target acquisition technique. Different positions and postures were used to decide the comfort level and the minimum distance to reach the targets. Twenty-four participants were hired in the second study, which aimed to understand the accuracy of control in target acquiring in VR with an eyes-free approach. The acquisition points were used for comparison with the actual points of the targets. The effects of the rotation of the user body on accuracy were also investigated. Based on the results of the first and second studies, the authors made improvements in the design of their eyes-free target acquisition system in VR. In the third user study, the improved eyes-free acquisition was compared with the eyes-engaged method with 16 participants. The acquisition performance and user experience were explored through an acquisition task with 18 targets. Different fields of view and the choice of a second task requiring visual focus were utilized on the comparison, that was based on a within-subject design. Overall, the eyes-free acquisition was preferred by most of the participants (i.e., 13 out of 16), especially when the field of view was small and when there was a second task. The eyes-free method had a higher acquisition speed, less fatigue and cybersickness, and less distraction although it could require more mental demand and result in lower accuracy. The future work was pointed out as investigating the eyes-free target acquisition with body or controller references.

1.7.8 Breathing-Based Input in VR

Advancements in VR yielded various unconventional forms of input in search of improved user experience. Very few previous works in VR had incorporated breathing in user experience and no previous works to date had utilized breathing as a form of input. Addressing this gap, recently, Sra et al. explored utilizing breathing signals as a direct controlling interface in VR games for improved immersiveness [78]. The authors designed four breathing actions and implemented two related VR games to explore the proposed input mechanism which was based on breathing. To evaluate the proposed system, they conducted a user study with 16 participants, to understand and suggest a better breath-based VR design. In the study, five design strategies that were proposed along with four breathing actions: gale, waft, gust, and calm. A Zephyr sensor was connected with the VR system to detect the users' inhalation and blowing actions. Those four breathing actions were each assigned to a corresponding power in the created VR games. To compare with the traditional hand-held controller input, the relative four areas on the controller trackpad were also assigned with the same respective power in the developed VR games. The researchers built the breathing actions into two

VR games: a single-player first-person shooting (FPS) game and a two-player ball game. In these games, different actions of breathing represented different "superpower" actions. For example, in the FPS game, the gale induced fire-causing bigger damage to the enemies. Within-subjects design was used in the user study to compare the breath input mechanism with the conventional controller-based input. Each participant played each game four times, with and without the breathing input. The results showed that the breathing-based VR games gave the participants a better sense of presence, more fun and increased challenges. Based on their study, the authors provided design suggestions for future breath-based VR designs. Future research directions were suggested as exploring the effects of action mappings on the dynamics of direct physiological signals in VR games, comparing performance in breathing-based input versus non-breathing input in VR games.

1.7.9 360-Degrees Browsing in VR

Virtual and augmented reality is expected to take place of mobile devices by presenting the users' practical tools to perform everyday tasks, such as e-mails and browsing. Recently, user experience in browsing in VR has been gaining the attention of researchers. It was found that on an average, users spend at least 57.4% time switching tabs, which is heavily dependent on the size of the screen [79, 80]. To remedy this time loss, parallel web browsing may be utilized in VR, because of the advanced interaction and visualization techniques [81]. To address this, a recent notable study was performed by Toyoma et al. [82], in which the authors created a VR web browser, named VRowser. The researchers used the VRowser to provide webpage content comparisons, allocation, grouping, and retrieval. They kept the metaphors of web browsing while taking advantage of well-established interaction and visualization techniques in VR. To evaluate the effectiveness of the proposed system, the authors performed a user study. In the system, the immersive VR mode gave an unlimited screen for users and broke the restriction of screen size in conventional computer displays. 3D UI was also incorporated to improve user experience in web browsing. Suitable segregated interaction techniques were used to perform web-related and 3D tasks from the literature [83]. The authors created a VR Web Workspace for users to navigate around in a geographic layout to group and transfer webpages. The Web Page Management was created to perform 3D web-related tasks, such as selection, position, rotation, and scaling. The Web Tasks were utilized for input and interaction with the web content. HTC Vive controller was used for the interaction. Ten college students participated in the study which aimed to explore user satisfaction and webpage placement strategies in two VRWs with different geographical properties. Ten types of webpages were explored to measure the users' preference. A questionnaire and a semi-structured interview were conducted at the end of the study sessions. Data were collected from the identification of the location of each webpage and the user study. The results implied that the landmarks were mostly used to place and retrieve webpages. The locomotion techniques in the prototype were demonstrated to not be fast enough for smooth switching. The authors pointed out that further exploration was needed to ensure the creation of efficient virtual workspaces for 360-browsing in VR.

1.8 Conclusion

As the recent previous studies demonstrated, there are several application areas where VR can be beneficial, even to the point of eliminating the boundaries of what is possible to do in real life (e.g., talking with virtual future self in VR). However, the research in this recently emerging medium is still in its early stages and several future research questions and unexplored application areas remain untackled.

References

[1] Cheng, A., Yang, L. and Andersen, E. (2017). Teaching language and culture with a virtual reality game. In Proceedings of the 2017 CHI Conference on Human Factors in Computing Systems (pp. 541–549). ACM.

[2] Culbertson, G., Andersen, E., White, W., Zhang, D. and Jung, M. (2016). Crystallize: An immersive, collaborative game for second language learning. In Proceedings of the 19th ACM Conference on Computer-Supported Cooperative Work & Social Computing (pp. 636–647). ACM.

[3] Agarwal, S., Furukawa, Y., Snavely, N., Simon, I., Curless, B., Seitz, S.M. and Szeliski, R. (2011). Building rome in a day. Communications of the ACM, 54(10), pp. 105–112.

[4] Dudschig, C., de la Vega, I. and Kaup, B. (2014). Embodiment and second-language: Automatic activation of motor responses during

processing spatially associated L2 words and emotion L2 words in a vertical Stroop paradigm. Brain and Language, 132, pp. 14–21.

[5] Choo, L.B., Lin, D.T.A. and Pandian, A. (2012). Language learning approaches: A review of research on explicit and implicit learning in vocabulary acquisition. Procedia-Social and Behavioral Sciences, 55, pp. 852–860.

[6] Graham, S., Santos, D. and Francis-Brophy, E. (2014). Teacher beliefs about listening in a foreign language. Teaching and Teacher Education, 40, pp. 44–60.

[7] Vázquez, C., Xia, L., Aikawa, T. and Maes, P. (2018). Words in Motion: Kinesthetic Language Learning in Virtual Reality. In 2018 IEEE 18th International Conference on Advanced Learning Technologies (ICALT) (pp. 272–276). IEEE.

[8] Freedman, R.A. (1996). Challenges in teaching and learning introductory physics. In from High-Temperature Superconductivity to Microminiature Refrigeration (pp. 313–322). Springer, Boston, MA.

[9] Grivokostopoulou, F., Perikos, I., Kovas, K. and Hatzilygeroudis, I. (2016). Learning approaches in a 3D virtual environment for learning energy generation from renewable sources. In The Twenty-Ninth International Flairs Conference.

[10] Ma, T., Xiao, X., Wee, W., Han, C.Y. and Zhou, X. (2014). A 3D virtual learning system for STEM education. In International Conference on Virtual, Augmented and Mixed Reality (pp. 63–72). Springer, Cham.

[11] Jimoyiannis, A. and Komis, V. (2001). Computer simulations in physics teaching and learning: a case study on students' understanding of trajectory motion. Computers & Education, 36(2), pp. 183–204.

[12] Saito, S., Hirota, K. and Nojima, T. (2017). KKse: Safety education system of the child in the kitchen knife cooking. In 2017 IEEE Virtual Reality (VR) (pp. 321–322). IEEE.

[13] Freeman, D., Reeve, S., Robinson, A., Ehlers, A., Clark, D., Spanlang, B. and Slater, M. (2017). Virtual reality in the assessment, understanding, and treatment of mental health disorders. Psychological Medicine, 47(14), pp. 2393–2400.

[14] Martono, N.P., Yamaguchi, T. and Ohwada, H. (2016). Utilizing finger movement data to cluster patients with everyday action impairment. In 2016 IEEE 15th International Conference on Cognitive Informatics & Cognitive Computing (ICCI* CC) (pp. 459–464). IEEE.

[15] Drzezga, A., Grimmer, T., Peller, M., Wermke, M., Siebner, H., Rauschecker, J.P., Schwaiger, M. and Kurz, A. (2005). Impaired cross-modal inhibition in Alzheimer disease. PLoS Medicine, 2(10), p. e288.

[16] Wiederhold, B.K., Riva, G. and Gutiérrez-Maldonado, J. (2016). Virtual reality in the assessment and treatment of weight-related disorders. Cyberpsychology, Behavior, and Social Networking, 19(2), pp. 67–73.

[17] Thomas, J.G. and Bond, D.S. (2014). Review of innovations in digital health technology to promote weight control. Current Diabetes Reports, 14(5), p. 485.

[18] Lafond, E., Riva, G., Gutierrez-Maldonado, J. and Wiederhold, B.K. (2016). Eating disorders and obesity in virtual reality: a comprehensive research chart. Cyberpsychology, Behavior, and Social Networking, 19(2), pp. 141–147.

[19] Perpiñá, C. and Roncero, M. (2016). Similarities and differences between eating disorders and obese patients in a virtual environment for normalizing eating patterns. Comprehensive Psychiatry, 67, pp. 39–45.

[20] de Carvalho, M., Dias, T., Duchesne, M., Nardi, A. and Appolinario, J. (2017). Virtual reality as a promising strategy in the assessment and treatment of bulimia nervosa and binge eating disorder: a systematic review. Behavioral Sciences, 7(3), p. 43.

[21] Keizer, A., van Elburg, A., Helms, R. and Dijkerman, H.C. (2016). A virtual reality full body illusion improves body image disturbance in anorexia nervosa. PloS One, 11(10), p. e0163921.

[22] Mountford, V.A., Tchanturia, K. and Valmaggia, L. (2016). "What Are You Thinking When You Look at Me?" A Pilot Study of the Use of Virtual Reality in Body Image. Cyberpsychology, Behavior, and Social Networking, 19(2), pp. 93–99.

[23] Ferrer-García, M. and Gutiérrez-Maldonado, J. (2012). The use of virtual reality in the treatment of eating disorders.

[24] Pla-Sanjuanelo, J., Ferrer-García, M., Vilalta-Abella, F., Riva, G., Dakanalis, A., Ribas-Sabaté, J., Andreu-Gracia, A., Fernandez-Aranda, F., Sanchez-Diaz, I., Escandón-Nagel, N. and Gomez-Tricio, O. (2017). Testing virtual reality-based cue-exposure software: Which cue-elicited responses best discriminate between patients with eating disorders and healthy controls?. Eating and Weight Disorders-Studies on Anorexia, Bulimia and Obesity, pp. 1–9.

[25] Garner, D.M. (2004). EDI-3, eating disorder inventory-3: Professional manual. Psychological Assessment Resources, Incorporated.

[26] Pla-Sanjuanelo, J., Ferrer-Garcia, M., Vilalta-Abella, F., Riva, G., Dakanalis, A., Ribas-Sabaté, J., Andreu-Gracia, A., Fernandez-Aranda, F., Sánchez, I., Escandón-Nagel, N. and Gomez-Tricio, O. (2017). VR-based cue-exposure therapy (VR-CET) versus VR-CET plus pharmacotherapy in the treatment of bulimic-type eating disorders.

[27] Paslakis, G., Fauck, V., Röder, K., Rauh, E., Rauh, M. and Erim, Y. (2017). Virtual reality jogging as a novel exposure paradigm for the acute urge to be physically active in patients with eating disorders: Implications for treatment. International Journal of Eating Disorders, 50(11), pp. 1243–1246.

[28] Keyes, A., Woerwag Mehta, S., Bartholdy, S., Koskina, A., Middleton, B., Connan, F., Webster, P., Schmidt, U. and Campbell, I.C. (2015). Physical activity and the drive to exercise in anorexia nervosa. International Journal of Eating Disorders, 48(1), pp. 46–54.

[29] Ferrer-Garcia, M., Pla-Sanjuanelo, J., Dakanalis, A., Vilalta-Abella, F., Riva, G., Fernandez-Aranda, F., Forcano, L., Riesco, N., Sánchez, I., Clerici, M. and Ribas-Sabaté, J. (2019). A randomized trial of virtual reality-based cue exposure second-level therapy and cognitive behavior second-level therapy for bulimia nervosa and binge-eating disorder: outcome at six-month followup. Cyberpsychology, Behavior, and Social Networking, 22(1), pp. 60–68.

[30] Ferrer-García M., Gutiérrez-Maldonado J., Pla-Sanjuanelo J., et al. (2017). A randomised controlled comparison of second-level treatment approaches for treatment-resistant adults with bulimia nervosa and binge eating disorder: assessing the benefits of virtual reality cue exposure therapy. European Eating Disorders Review, 25, pp. 479–490.

[31] United States Department of Agriculture. Analyses of What We Eat in America, National Health and Nutrition Examination Survey (NHANES) data from 1999–2000 through 2009–2010. Last Accessed on September 29, 2019. Available at https://www.cdc.gov/nchs/nhanes/wweia.htm

[32] Oh, S.Y., Bailenson, J., Weisz, E. and Zaki, J. (2016). Virtually old: Embodied perspective taking and the reduction of ageism under threat. Computers in Human Behavior, 60, pp. 398–410.

[33] Aymerich-Franch, L. and Bailenson, J. (2014). The use of doppelgangers in virtual reality to treat public speaking anxiety: a gender comparison. In Proceedings of the International Society for Presence Research Annual Conference (pp. 173–186).

[34] Ahn, S.J., Le, A.M.T. and Bailenson, J. (2013). The effect of embodied experiences on self-other merging, attitude, and helping behavior. Media Psychology, 16(1), pp. 7–38.

[35] Fox, J. and Bailenson, J.N. (2010). The use of doppelgängers to promote health behavior change. CyberTherapy & Rehabilitation, 3(2), pp. 16–17.

[36] Ruiz, J.G., Andrade, A.D., Anam, R., Aguiar, R., Sun, H. and Roos, B.A. (2012). Using anthropomorphic avatars resembling sedentary older individuals as models to enhance self-efficacy and adherence to physical activity: psychophysiological correlates. Studies in Health Technology and Informatics, 173, pp. 405–411.

[37] Hershfield, H.E., Goldstein, D.G., Sharpe, W.F., Fox, J., Yeykelis, L., Carstensen, L.L. and Bailenson, J.N. (2011). Increasing saving behavior through age-progressed renderings of the future self. Journal of Marketing Research, 48(SPL), pp. S23–S37.

[38] Aymerich-Franch, L., Kizilcec, R.F. and Bailenson, J.N. (2014). The relationship between virtual self similarity and social anxiety. Frontiers in Human Neuroscience, 8, p. 944.

[39] Bailenson, J.N. (2012). Doppelgangers-a new form of self?. Psychologist, 25(1), pp. 36–38.

[40] Fox, J., Bailenson, J.N. and Ricciardi, T. (2012). Physiological responses to virtual selves and virtual others. Journal of CyberTherapy & Rehabilitation, 5(1), pp. 69–73.

[41] Aymerich-Franch, L., Karutz, C. and Bailenson, J.N. (2012). Effects of facial and voice similarity on presence in a public speaking virtual environment. In Proceedings of the International Society for Presence Research Annual Conference (pp. 24–26).

[42] Bandura, A. (2001). Social cognitive theory: An agentic perspective. Annual Review of Psychology, 52(1), pp. 1–26.

[43] Parfit, D. (1971). Personal identity. The Philosophical Review, 80(1), pp. 3–27.

[44] Schelling, T.C. (1984). Self-command in practice, in policy, and in a theory of rational choice. The American Economic Review, 74(2), pp. 1–11.

[45] Wilson, T.D. and Gilbert, D.T. (2005). Affective forecasting: Knowing what to want. Current Directions in Psychological Science, 14(3), pp. 131–134.

[46] Loewenstein, G. (1996). Out of control: Visceral influences on behavior. Organizational Behavior and Human Decision Processes, 65(3), pp. 272–292.

[47] Ahn, S.J.G., Bailenson, J.N. and Park, D. (2014). Short-and long-term effects of embodied experiences in immersive virtual environments on environmental locus of control and behavior. Computers in Human Behavior, 39, pp. 235–245.

[48] Polanin, J.R., Espelage, D.L. and Pigott, T.D. (2012). A Meta-Analysis of School-Based Bullying Prevention Programs' Effects on Bystander Intervention Behavior. School Psychology Review, 41(1).

[49] McEvoy, K.A., Oyekoya, O., Ivory, A.H. and Ivory, J.D. (2016). Through the eyes of a bystander: The promise and challenges of VR as a bullying prevention tool. In 2016 IEEE Virtual Reality (VR) (pp. 229–230). IEEE.

[50] Hock, P., Benedikter, S., Gugenheimer, J. and Rukzio, E. (2017). Carvr: Enabling in-car virtual reality entertainment. In Proceedings of the 2017 CHI Conference on Human Factors in Computing Systems (pp. 4034–4044). ACM.

[51] Slater, M., McCarthy, J. and Maringelli, F. (1998). The influence of body movement on subjective presence in virtual environments. Human Factors, 40(3), pp. 469–477.

[52] Van Kerrebroeck, H., Brengman, M. and Willems, K. (2017). Escaping the crowd: An experimental study on the impact of a Virtual Reality experience in a shopping mall. Computers in Human Behavior, 77, pp. 437–450.

[53] Pantano, E. and Timmermans, H. (2014). What is smart for retailing?. Procedia Environmental Sciences, 22, pp. 101–107.

[54] Renko, S. and Druzijanic, M. (2014). Perceived usefulness of innovative technology in retailing: Consumers.101-107.xperience in a shopping mall. Computtailing and Consumer Services, 21(5), pp. 836–843.

[55] Yee, N. (2006). The demographics, motivations, and derived experiences of users of massively multi-user online graphical environments. Presence: Teleoperators and Virtual Environments, 15(3), pp. 309–329.

[56] Kang, H., Lee, G., Kwon, S., Kwon, O., Kim, S. and Han, J. (2018). Flotation simulation in a cable-driven virtual environment – A study with parasailing. In Proceedings of the 2018 CHI Conference on Human Factors in Computing Systems (p. 632). ACM.

[57] Biocca, F. and Levy, M.R. (2013). Communication in the age of virtual reality. Routledge.

[58] Lee, J., Kim, B., Suh, B. and Koh, E. (2016). Exploring the front touch interface for virtual reality headsets. In Proceedings of the 2016 Chi Conference Extended Abstracts on Human Factors in Computing Systems (pp. 2585–2591). ACM.

[59] Heo, S., Chung, C., Lee, G. and Wigdor, D. (2018). Thor's hammer: An ungrounded force feedback device utilizing propeller-induced propulsive force. In Proceedings of the 2018 CHI Conference on Human Factors in Computing Systems (p. 525). ACM.

[60] McGill, M., Boland, D., Murray-Smith, R. and Brewster, S. (2015). A dose of reality: Overcoming usability challenges in VR head-mounted displays. In Proceedings of the 33rd Annual ACM Conference on Human Factors in Computing Systems (pp. 2143–2152). ACM.

[61] Van Wyk, E. and De Villiers, R. (2009). Virtual reality training applications for the mining industry. In Proceedings of the 6th International Conference on Computer Graphics, Virtual Reality, Visualisation and Interaction in Africa (pp. 53–63). ACM.

[62] McGill, M., Ng, A. and Brewster, S. (2017). I am the passenger: how visual motion cues can influence sickness for in-car VR. In Proceedings of the 2017 Chi Conference on Human Factors in Computing Systems (pp. 5655–5668). ACM.

[63] Knierim, P., Schwind, V., Feit, A.M., Nieuwenhuizen, F. and Henze, N. (2018). Physical keyboards in virtual reality: Analysis of typing performance and effects of avatar hands. In Proceedings of the 2018 CHI Conference on Human Factors in Computing Systems (p. 345). ACM.

[64] Hart, S.G. (2006). NASA-task load index (NASA-TLX); 20 years later. In Proceedings of the Human Factors and Ergonomics Society Annual Meeting (Vol. 50, No. 9, pp. 904–908). Sage CA: Los Angeles, CA: Sage publications.

[65] Witmer, B.G. and Singer, M.J. (1998). Measuring presence in virtual environments: A presence questionnaire. Presence, 7(3), pp. 225–240.

[66] Whitmire, E., Benko, H., Holz, C., Ofek, E. and Sinclair, M. (2018). Haptic revolver: Touch, shear, texture, and shape rendering on a reconfigurable virtual reality controller. In Proceedings of the 2018 CHI Conference on Human Factors in Computing Systems (p. 86). ACM.

[67] Guiard, Y. (1987). Asymmetric division of labor in human skilled bimanual action: The kinematic chain as a model. Journal of Motor Behavior, 19(4), pp. 486–517.

[68] Strasnick, E., Holz, C., Ofek, E., Sinclair, M. and Benko, H. (2018). Haptic links: bimanual haptics for virtual reality using variable stiffness actuation. In Proceedings of the 2018 CHI Conference on Human Factors in Computing Systems (p. 644). ACM.

[69] Millais, P., Jones, S.L. and Kelly, R. (2018). Exploring data in virtual reality: Comparisons with 2d data visualizations. In Extended Abstracts of the 2018 CHI Conference on Human Factors in Computing Systems (p. LBW007). ACM.

[70] Chen, X., Self, J.Z., House, L. and North, C. (2016). Be the data: A new approach for Immersive analytics. In 2016 Workshop on Immersive Analytics (IA) (pp. 32–37). IEEE.

[71] Brunhart-Lupo, N., Bush, B.W., Gruchalla, K. and Smith, S. (2016). Simulation exploration through immersive parallel planes. In 2016 Workshop on Immersive Analytics (IA) (pp. 19–24). IEEE.

[72] Inselberg, A. (1985). The plane with parallel coordinates. The Visual Computer, 1(2), pp. 69–91.

[73] Saraiya, P., North, C. and Duca, K. (2005). An insight-based methodology for evaluating bioinformatics visualizations. IEEE Transactions on Visualization and Computer Graphics, 11(4), pp. 443–456.

[74] Edge, D. and Blackwell, A.F. (2009). Peripheral tangible interaction by analytic design. In Proceedings of the 3rd International Conference on Tangible and Embedded Interaction (pp. 69–76). ACM.

[75] Hausen, D., Tabard, A., Von Thermann, A., Holzner, K. and Butz, A. (2014). Evaluating peripheral interaction. In Proceedings of the 8th International Conference on Tangible, Embedded and Embodied Interaction (pp. 21–28). ACM.

[76] Xiao, R. and Benko, H. (2016). Augmenting the field-of-view of head-mounted displays with sparse peripheral displays. In Proceedings of the 2016 CHI Conference on Human Factors in Computing Systems (pp. 1221–1232). ACM.

[77] Yan, Y., Yu, C., Ma, X., Huang, S., Iqbal, H. and Shi, Y. (2018). Eyes-Free Target Acquisition in Interaction Space around the Body for Virtual Reality. In Proceedings of the 2018 CHI Conference on Human Factors in Computing Systems (p. 42). ACM.

[78] Sra, M., Xu, X. and Maes, P. (2018). Breathvr: Leveraging breathing as a directly controlled interface for virtual reality games. In Proceedings of the 2018 CHI Conference on Human Factors in Computing Systems (p. 340). ACM.

[79] Dubroy, P. and Balakrishnan, R. (2010). A study of tabbed browsing among mozilla firefox users. In Proceedings of the SIGCHI Conference on Human Factors in Computing Systems (pp. 673–682). ACM.

[80] Xu, W., Yu, C., Zhao, S., Liu, J. and Shi, Y. (2013). Facilitating parallel web browsing through multiple-page view. In Proceedings of the SIGCHI Conference on Human Factors in Computing Systems (pp. 2167–2170). ACM.

[81] LaViola Jr., J.J., Kruijff, E., McMahan, R.P., Bowman, D. and Poupyrev, I.P. (2017). 3D user interfaces: theory and practice. Addison-Wesley Professional.

[82] Toyama, S., Al Sada, M. and Nakajima, T. (2018). VRowser: A Virtual Reality Parallel Web Browser. In International Conference on Virtual, Augmented and Mixed Reality (pp. 230–244). Springer, Cham.

[83] Bowman, D.A., Hodges, L.F. and Bolter, J. (1998). The virtual venue: User-computer interaction in information-rich virtual environments. Presence, 7(5), pp. 478–493.

2

Digital and Visual Literacy, Video Games, and Virtual Reality

Jack Clark

School of Information, University of Arizona, USA
E-mail: jaclarkent@email.arizona.edu

In contemporary society, global digital interaction is more accessible in leisure activities than in educational contexts. Children are interacting with one another in digital arenas like social media, fan fiction, and video games; taking personal responsibility for their learning. They are learning the skills needed to thrive in this 21st century through their leisure activities and online gamified contexts, as opposed to academically in the classroom. There are socioeconomic divides in all facets of learning including the development of digital and visual literacies, visual stimulation, and immersive interaction. Unfortunately, most individuals are unaware of the affordances for learning through video games, let alone virtual reality (VR) settings.

This chapter will discuss the need to research and improve visual and digital literacies in order to understand the effects of video games and VR learning in academic settings. First, the lineage of visual literacy growth will be accessed in light of important literature and advancements of technology. Second, current research on video gamified learning will be mapped onto the potential uses for classroom education. Third, this paper will discuss the current benefits and applications of VR and immersive learning in contemporary culture. Finally, the paper will conclude with an educated interpretation of the possible future interactions of VR learning in both leisure activities and academic settings. The aim of this paper was to create an encapsulation of gamified and immersive learning in its current state; with an explanation of how video games and VR allow humans to access experiences previously thought impossible.

2.1 Introduction

Recent figures from the U.S suggest the total market value of the video game industry in 2014 to be $22.4 billion [1]. Video games are ubiquitous, with four out of five households owning a device used for this digital interaction [1]. This progression of technology in today's society has led to stark differentiating perspectives on academic versus leisure activities. Throughout history, visual stimulation is more common in leisure activities than in education. Education has been rooted in reading and writing, with art analysis in the background. However, leisure activities today focus more on visual stimulation with minimal reading or very little writing. Children are developing their visual and digital literacies in these leisure activities through online gamified contexts, compared to academically in the classroom. This occurrence is causing cultural and socioeconomic divide in visual stimulation and development of visual and digital literacies [2]. Unfortunately, most individuals are unaware of the affordances for learning in digital settings, especially through video game arenas or VR contexts. Even with the accessibility of communication being digitally exploited there is a lack of focus in becoming visually and digitally literate in American public schools.

There is a growing body of research and literature centered on digital tools and the dichotomous nature of leisure and academic activities. While not widely accepted, video games, VR, and many other out-of-classroom digital interactions can be utilized for academic purposes and the development of 21st-century skills [3]. It has been found that learning through video games can have positive effects on attention and focus [4]. Video game players, compared to non-video game players, show a greater ability to focus on multiple moving items at once [4]. Coincidentally, playing all types of video games improves attention and visual working memory, which in turn can have educational benefits. These benefits could possibly include quicker receptions of visual information and improvement of visual and digital literacies. Unfortunately, not all games are accessible to all kinds of people and many are not structured appropriately for educational use [5]. Lack of accessibility of these learning experiences is increasing the achievement gap and "digital divide" for digital and visual capabilities [5]. Individuals with language barriers or sensory impairments will not be able to interact with technological experience in the same manner an individual without barriers or impairments would.

Children in higher socioeconomic status households will have more exposure to this kind of visual development compared to children growing up in poverty [5]. The capitalistic nature of the video game industry has increased expenses needed to stay on the cutting edge of technological interaction, creating a divide in digital and visual literacies between groups of people and different households [5]. More important, only 5% of video game purchases are made primarily for educational purposes [1]. In my experience working at the Boys and Girls Club in Tucson, Arizona with an average of 100 attendees daily for 8 months in 2016, children believe educational video games are "boring" and unattractive compared to non-educational video games like Call of Duty, Skyrim, or Grand Theft Auto.

Since the release of Carmen San Diego in 1987, children, parents, educators, and game distributors have not agreed on a balance between accessibility, enjoyment, and content within educational video games [6]. This disagreement has led to advocacy for structures designed specifically to foster learning and accessibility for all kinds of people. To better understand this phenomenon, this chapter is divided into three main sections: (1) the growth of visual literacy, (2) video game learning, and (3) VR. The goal is to highlight the necessity for including video games and VR in the educational learning process, in order to help narrow the achievement gap and digital divide between the groups of people and diverse students [5].

2.2 The Growth of Visual Literacy

The concept of thinking visually is something most individuals are familiar with. The use of graphs to depict data, visually representing emotions, or sharing artistic skills are considered traits of visual thinkers [7]. The exploitation of this concept is not new to pedagogical practices, however, it has yet to be applied to the innovative technologies of the 21st century. In today's world we use visual skills more intensely than ever before, especially in contemporary leisure activities like reading graphic novels, playing video games, and interacting with VR platforms. The rapid growth of technological devices, video games, and computer-generated imaging (CGI) has heightened our visual capabilities and changed the way we interpret visual information. Now more than ever, it is crucial to develop an understanding of what it means to be visually literate and how this can be taught in education.

The following section will define the concept of visual literacy by providing a basis for terminology, along with rooting the idea in known literature. It will focus on what it means to think visually and how this concept can be trans-mediated into speech or written word. Finally, this section will touch on why there should be a blending of technologies used in today's leisure activities and classroom education.

2.3 Thinking in Pictures

One of the key researchers and activists for conceptualizing thinking in pictures is Temple Grandin. Temple is considered an autistic savant, with the drive to understand how people think and apply her research towards bettering the education of today's youth. She argues there are three classifications of cognition: visual, pattern, and verbal thinkers [7]; each bringing a unique set of skills that can be used in collective intelligence [8]. Temple Grandin is known for her unique visual cognition and how she used it to analyze emotional change in cattle being raised for human consumption. In her research, the books, and movies surrounding Temple's career, she defines her visual thinking with specific terminology. The effective use of this terminology is the missing link for transmediating the concept of visual literacy to a learned skill.

In Temple's work, she explains the visual interpretations of her thinking in the same terminology as an art analyst. Much like the analysis of a Monet, she uses words like contrast, color, focus, angle, frame, brightness, saturation, etc. to explain the visual stimulations she is receiving from the environment [6]. Continuing like an esteemed art critic, Temple applies this terminology to the interpretation of various visual elements and their interactions with one another [6]. For example, the color of a flag as it is swirled by the wind causing stress or balking in cows. Temple used these stimulations and interpretations to discover how specific visual cues elicit certain responses [6]. This would imply communication of information using only images, changing how we conceptualize cognitive experiences.

After dissecting Temple's work, it is clear that there is a direct connection to the visual terminology which exists in art and analysis. However, with any analysis of artwork and creative content, there is a silver lining of subjectivity. Each individual gains something different from an image and can gain something new each time they view that image. This subjectivity makes it difficult to directly show the connection of a specific visual cue to a specific elicited behavior. For this reason, using eye trackers to show a connection to a behavior is incomplete [9]. Nonetheless, researchers have been fascinated with the communication through visual stimulation; blossoming fields like advertisement analysis, text and visual interactions, film studies, reality studies, human-computer interaction, and multimodal research. The focus is to connect these visual stimulations and analyze communication of ideas, emotions, and to understand its occurrence in leisure activities. Through heightened visual stimulations, learners received through technological devices in the 21st century, understanding this type of communication is crucial to the concept of visual literacy.

2.4 Comic Books and Film Studies

One of the steppingstones in Temple's terminology and research into multimodal interactions such as video games and VR are comic books. In the beginning of comic creation, these books were perceived only for children, without any regard for the positive affordances they were offering to cognitive development and visual meaning-making. It was not until the naming of The Watchmen by Alan Moore, Dave Gibbons, and John Higgins, in the Time magazine's top 100 best novels of the 20th century that comic books were given the equal prestige of written novels. This acceptance cannot be credited to one book or the winning of a single competition, yet the creation of an objective structure for analyzing text and visual components in one location [10]. The comic book analysis is the first blend of leisure activities and education. This art form blends text and images in one emotional and thought-provoking arena. Comic books and graphic novels offer a stagnant context to dissecting specific visual cues and the emotions they elicit.

One of the reasons a blending of leisure activities and education can be achieved is because of the terminology being used to describe the cognitive occurrences in these visual interactions. In the same fashion as shown by Temple Grandin and accredited by art analysts to explain and interpret comic book visual cues, words like contrast, color, brightness, angle, frame, focus, etc. are used [7, 10]. For example, the term contrast is rooted in opposition and is universally seen as an objective concept. Contrast between lighting spectrums, color choices, or conceptual or artistic ideas is where the subjective variability of opposition occurs. Moreover, imagine a picture of a sunken ship and a beautiful sunrise. Some individuals will comment on the contrasting colors of the orange sun and the blue sea, others will comment the sea is dark and the sun is bright, or some may be more conceptual, focusing on the contrast between death in a sunken ship and rebirth in a sunrise. Utilizing the same terminology in a variety of contexts begins to solidify and objectify the definition of each term being used. The more objective and universally understood the terminology is, the more likely there is consensus of interpretation and foundation for the requirements of being visually literate.

Another arena where these terminology definitions are considered universal and regularly used is in film studies. As in comic book analysis, film analysts use words like frame, angle, contrast, color, lighting, etc. The differences between comic book and film analysis is the inclusion of movement and type multimodal interaction. Adding movement adds visual stimulation and changes the way information is communicated. In a comic book, the mind fills in frame gaps (gutters) with the movement of characters yet they are not physically moving on the page in front of them [10]. This allows for variability in perception and interpretation of the material [10]. Film depicts the movement for the viewer, creating a set display of the character's interaction. For example, a comic book can have two consecutive frames, one of a female in the dining room and the other of her in the kitchen. The mind creates the character walking from the dining room to the kitchen. In a film, this movement is shown, allowing an analysis of the walking itself. Questions regarding the emotionality of the walk could be answered quicker and be more readily seen as an objective experience in a film than in a comic book from the viewer's perspective. That being said, the creator of each respective work holds the discretion to include or exclude any imagery or parts of the storyline they choose. For example, a comic may include text in the frame depicting movement like 'stomp' near her feet, whereas a movie may show the woman stomping her feet as she enters the kitchen.

The important distinction between comic books and film is the way they differ in sensory stimulation. Film creates an auditory and visual combination in one experience, whereas comic books utilize written word and imagery. Both are multimodal contexts, yet they drive stimulation very differently. Further, these different experiences are not accessible to all kinds of people. Visually impaired individuals will gain much less from a comic than they will from a film because of the auditory interaction. This would also hold true for an individual with hearing impairments; they will be able to interact with a comic book much like someone without a hearing impairment, however, they will lose part of the experience if the content is solely visually interpreted. Yet, the utilization of the same terminology in different contexts supports an objective understanding of what it means to be visually literate. Following the same concept applied previously, the use of the same terminology in different contexts, helps refine the definition of being visually literate term and its meaning universally.

Solidifying universal objective definitions regarding visual cues creates a framework for teaching visual interpretations and increasing visual literacy in education. It is possible to come to an understanding of the definitions themselves as well as design interactive projects analyzing both comic and film works. Further, the similarity in terminology allows for the comparison of comic books and films regarding the same content. For example, a possible project could be understanding the emotional interpretation differences between The Watchmen graphic novel and The Watchmen movie. This idea also helps blend leisure and academic activities while supporting educational interaction.

2.5 Video Games

Being visually literate is no longer simply being able to interpret the world and human experience through visual works and define points of sensory interaction, it has progressed to being able to control the various sensory stimulations and interpretations in any given environment. Education is just beginning to include the concept of controlling multimodal sensory interactions and expand the understanding of visual literacy. The best example of control in a multimodal situation is video games. Almost all video games, both commercial and educational involve multiple sensory stimulation and incorporate the interaction of text, image, sounds, and movement in one location [46]. This being the case, the same terminology in both comic and film studies could be applied to understanding the stimulation of video games.

The fascinating aspect of video games is the user's ability to control the experience itself. Adding this layer of control takes the experience from observational to participatory, further engaging cognitive interaction with the stimulations being used. This context fosters motivation, collaboration, self-regulation, and depicts the learner's perspective while exhibiting the emotional regulation of the individual [46]. With the rapid growth of technology and the video game industry progressing in leisure activities, what was previously believed to be fictitious forms of learning are now obtainable. The video game arena is making the impossible, possible through media and multimodal stimulation. In a digital age that is networked and interwoven, learning is no longer restricted to classroom settings. This drives the necessity for understanding how visual stimulation, control, and learning are occurring in video games and to explore what their application could be in academic settings.

2.6 Video Game Learning

The structure of video games offers many benefits to the player while being a favorable context for understanding multimodal learning and intrinsic motivation [46]. Regardless of the video game type, there is a positive relationship between video gameplay and improved attention [4]. Also, the immediate feedback embedded within video game structure allows the opportunity for players to monitor and evaluate their own learning while complex problem solving [11–13]. This, combined with the autonomous structure of video-game learning, offers an affordance for self-regulated learning [14–16]. Continuously, role-playing and MMORPGs structure provides the opportunity for player collaboration and communication while furthering gameplay [14, 16, 46]. MMORPG's structure specifically allows players to be autonomous within an interactive environment [14, 16]. This type of open communication will foster collaborative independent learning and provide a potential arena that is beneficial for academic use.

2.7 Pretend Play and Situated Learning

Many educators view 'play' as a beneficial activity to foster development [17]. During play, individuals observe and collaborate towards a common goal, while thinking creatively to solve complex problems [18]. One type of play is imaginative (pretend play), with character roles and emulations of current world issues [19]. For this type of activity to be considered pretend play, it must include a mental representation different from reality, a projection of that mental representation onto reality, and the awareness of the representation, as well as its projection [20]. These three components allow participants to include their imagination with their current environment while engaging in a leisure activity [20]. A non-digital example is using a banana for a telephone.

Similarly, situated learning is the placement of the learner in a particular context to drive social, cultural, and academic interaction [21]. Applying these parameters to the content being learned allows the learner to see how the information being received is utilized in social settings [21]. This is beneficial to learning because it allows the participant to think critically about the information used in each context and how it appropriately interacts with the respective situation [21, 22]. It is clear this context of situated learning is embedded in most games, both commercial and educational, available on the market today. Within the video game industry alone, there is a spectrum of non-immersive virtually situated contexts ranging from full storyline narratives to a few details of the situated scenario. Games without any situated learning context include mini-games or module style games like Candy Crush or Poker.

Often, pretend play and situated learning are viewed as more beneficial for younger participants, however, the benefits can extend to adults if structured appropriately [21, 23]. For instance, despite common misconceptions, there are measurable cognitive benefits for play and engaging with video games at all ages [4, 15, 24, 25]. Specifically, role-playing video games allow users to positively engage in pretend play, situated learning, and collaborative interactions [15, 21]. Both pretend play and situated learning show positive effects on creative capacities and fosters innovative, agentive thinking when utilized in video games [15, 21, 26]. Video game platforms provide an arena for participants to observe, interact, and engage with tools beneficial to the overall learning process [11, 27–29].

2.8 Collaboration

Through this type of play and situated learning, video game players benefit from the opportunity to collaborate with other individuals [11, 15, 46]. Much like physically working with someone to complete a common goal, video games and the video game community foster communication between players [28, 29]; see also [11, 27, 46]. For games like Assassin's Creed (one-player, narrative role-playing game), players do not directly interact with each other as they do in massively multiplayer online role-playing games (MMORPGs) [29]. However, communication is fostered through online chat forums and discussion boards [28, 29]. In MMORPGs, players must collaborate and compete to propel gameplay and further game achievement [29]. From a pedagogical perspective, video game platforms can benefit all types of learners, especially individuals with language barriers, social, or physical needs different from the stereotypical player or "the gamer" (i.e. Autism Spectrum Disorder) [12]. Moreover, in MMORPGs like World of Warcraft, a player's age ranges from youth to adult and come from all over the world. This diversity allows for the virtual interaction between populations while fostering the opportunity for communication and possible language learning [12].

The opportunity for communication in video games and virtual worlds aids inter-player collaboration and allows for the exchange of multiple perspectives, yet also promotes competition. The nature of the immediate statistical feedback within the structures of video games usually fosters more competition than collaboration, especially in popular commercial games [15]. Most currently available video games log every aspect of the player's game usage, including time taken to complete tasks, activity during tasks, and order of multiple task completion [29, 30]. These statistics are scored by the game and placed into an overall player ranking available online to other players [29, 30]. From a pedagogical perspective, these statistics create a collection of data encompassing individual learning at arms-length from the interaction of collaboration [46]. The researcher can see how the individual is learning from the experience, simultaneous to interpreting their interaction with other players. More importantly, understanding the differences between collaboration and competition will provide insights to group decision making, group strategic planning, and group goal achievement.

2.9 Autonomy

Unfortunately, teachers in traditional classrooms require students to master certain skills before continuing to the next task. This type of learning is linear or vertical and not catered to the individual speed or needs of each student, but the overall needs of all students collectively [8]. By contrast, video games present a few different task options towards one goal at the same time [11, 16, 28, 31]. This forces the player to make thoughtful choices regarding the order of the task objectives and how they will be reached [28, 31]. This structure also provides scaffolding and hints for the learner through different stages of the game [31]. In these video games, players can cater their interests and instructional path to best fit their needs during gameplay [46]. As players make choices regarding order of tasks, they are also learning collaborative strategic planning for solving complex problems [16].

2.10 Self-Regulated Learning

Role-playing games emphasize autonomy for strategic planning, complex problem solving, and self-regulated learning [14–16, 45]. The player is presented with multiple task options at once, yet it is the player's choice of which to engage. This type of environment enables the players to self-regulate their gameplay and achievement. Self-regulated learning is defined as the metacognitive process of initiating, guiding, and comparing own performance to self-set standards [32, 33, 45]. This process begins with self-instruction, continues into self-monitoring/control, and leads to self-evaluation. This type of learning is dependent on combinations of situational parameters, self-set standardized, the learner's past experiences, and their choice of public versus private settings. The ability to control these combinations is aided by the structural advantages and design of video games.

According to Anderson [34], there are five components of self-regulating learning: prepping and planning, selection of learning strategy(s), monitoring strategy implementation, correlating and combining strategies, and evaluating the effectiveness of implementation. Specifically, first-person shooter RPGs like Call of Duty II and Elder Scrolls: Skyrim provides supplemental materials for the player, allowing them to be autonomous and self-regulate their own learning [44]. These materials include, but are not limited to, progress bars, in-game player statistics, arena maps, highlighted location destinations (storyline use), lists of objectives and goal setting, both proximal and distal goal setting, and collaboration opportunities with other players (both verbally and nonverbal). Continuous use of these supplemental materials and structural mechanisms allows the player to track and adjust their learning in a specifically personalized way. This process makes video game playing individualistic while offering an arena for emotional regulation within a leisure activity.

The affordability for self-regulated learning is one of the most beneficial factors of using video games as an instructional method [11–13]. Video games best highlight the self-regulation process within MMORPGs, however, this process is true for all role-playing video games whether they are multiplayer or not [11–13]. During gameplay, players command an avatar that is representative of the player. The player has control over the avatar's physical and cognitive actions, including driving their motivation. Once a task is instructed, players will monitor game performance and regulate the avatar's cognitive and physical activities until the task is complete [12,13]. Finally, the player will reflect on their overall task performance in conjunction with the avatar after the task is completed. Video games cannot ensure these judgments will be made accurately, yet the opportunity for learning is provided [12, 13].

The opportunity for self-regulated learning within video game structure is the reason they should be utilized as instructional tools. The process of self-regulated learning is metacognitive, yet fostered by role-playing games. In these types of games, players must seek a task/activity/objective to complete before receiving a reward (i.e. game currency) [27]. This provides the opportunity for players to regulate their learning and while showing intrinsic motivation through game continuation [35, 36]. Players can strategize and plan the execution of their tasks for efficiency and maximum reward based on personal interests. Moreover, players can monitor and evaluate the learning strategies used in current gameplay to further current and future game achievement [12, 13, 45].

2.11 Observational vs. Participatory Learning and Volitional Control

Whether video game playing is tied to observational learning in non-immersive reality settings is continuously being debated [34]. Watching a character or avatar commit action is observable to players without them physically engaging in that action [27]. However, player action is required for the avatar to engage in a task [27]. For example, in Sims, a single player role-playing game, players command their avatars to go swimming. The player is still sitting in a chair inside, whereas the avatar is swimming laps in the virtual pool. Moreover, control is still with players outside the game boundaries, yet action is within [27]. Having this type of control allows partial observational learning because the player is not immersively engaging physically with the task in their own reality. Yet, because players have volitional control over the avatar, direct learning can be illustrated [34].

Volitional control is the decision and action of committing to engage with a behavior or task [16]. Volitional control in role-playing video games can have both positive and negative effects on behavior (i.e. Call of Duty, Grand Theft Auto) [35, 36]. In the most recent analysis of violent video game effects, an increase in aggression scores and cognition, as well as a decrease in empathy and sensitization to violence were found [35]. This is due to both the individual and situation being the input attributions to a person's behavior [35]. These attributions create a temporary internal state that can be appraised, reviewed, and revised for self-regulated learning [35]. For example, volitional control in commercial first-person shooter games that promote violent imaging and constant death in their internal state will negatively affect behavior [44].

However, knowingly engaging with volitional control when learning presents a significant positive cognitive effect compared to automatic cognitive processes like skill acquisition and initial perception [14, 16]. In other words, the metacognitive awareness of committing the action has benefits in comparison to committing the action more automatically without metacognitive choice making. Having visual volitional control in video games positively aids learning, whereas physically having volitional control, like in physically immersive VR settings during violent role-playing scenarios might not [16]. During VR there is physical volitional control, the players feel they themselves are carrying out the actions of the avatar as they happen [16]. When the player fails to complete a task in VR settings, they allocate the error to themselves, whereas in non-immersive settings, players allocate some of the error to the game or avatar [16]. Therefore, it can be concluded visually controlling an avatar in role-playing games allows the player to feel disconnected from the avatar's actions. [16]. In some of the more violent role-playing games, it may be important to not associate those actions to reality. Yet, away from war-like scenarios, the benefits and positive affordances of VR learning outweigh the negative.

2.12 Current Benefits and Applications of Virtual Reality

The cognitive benefits of video game learning apply to VR settings as well. This physically engaging arena offers aspects of situated learning, collaboration between users, strategic planning, visual meaning-making, self-regulated learning, motivation, engagement, and enjoyment. The difference with VR is that the user is fully immersed in the context, without any information from the real world around them. The user wears a head-mounted device (HMD) that only allows them to interact with the virtual world and inhibits stimulation from the outside world (reality). This context helps the user have a deeper immersive experience with a situated learning setting and roots their actions in participation with their surroundings. As the learner engages with the immersive visualization, they are applying self-regulation techniques to complete each task [45]. Much like the real world, when an individual makes a mistake they consider why the error was made and begin to correct it for future, more effective application. Considering virtual arenas are newer and users have had less exposure with the device, many VR games are structured to give multiple opportunities to regulate the effectiveness of a particular strategy. This style of horizontal learning allows the learner to cater the experience to their personal needs for learning and monitor effectiveness of the tool along with game achievement.

The immersive experience of VR allows for a more realistic context than non-virtual video game interactions by allowing the user to learn with their entire body in a fully physically inclusive learning experience. This includes muscle movement memory, coordination, and unconscious reactions [8]. As seen in athletic training, the constant repetition of a movement helps create an internal muscle remembrance of that movement and makes it easier to commit that movement in the future. Much like constant repetition of vocabulary-definition study, VR could increase recall and retrieval of those movements. This is especially beneficial to an application of VR learning in work-related settings. For example, a mechanic may want to learn how to change a transmission but does not want to practice on a perfectly working car. With VR learning, the mechanic can develop muscle memory for changing a transmission before being placed in that situation in real-life settings.

This is also one of the reasons why VR is being used in the medical field to train medical students before operating on a cadaver. Cadavers are expensive and may not be readily available for medical students to practice. Yet, the students need practice to gain muscle memory of how to proceed with the surgery. While the virtual setting is going to be very different than a real cadaver, this arena allows for a steppingstone or checkpoint of knowledge and skill. Many health professionals will argue this difference is too great to apply similarity of a virtual setting to the real scenario. However, the use of virtual reality in medicine is not to replace practice on cadavers yet provide a scaffolded curriculum incorporating muscle memory before adding higher expenses and the more emotional stimulations of working with a real corpse.

The most important aspect of VR is its positive affordance for accessibility. VR is enabling a variety of content to be accessible to multiple kinds of people all over the globe. Archeologists are able to immerse themselves in caves or temples thousands of miles away from the comfort of their office. Language learners can practice with native speakers through digital avatars and virtual meeting spots. The average person can visit any museum or art exhibit in the world. And, psychologist can expose victims to their fears in safe, forgiving, and flexible environments. VR offers a more controlled version of the content, yet it allows for immersive interaction, nonetheless.

2.13 Virtual Reality in the Future

VR does more than allow accessibility of a variety of content, it allows all kinds of individuals to have access to information they were never capable of before. The latest technological advancement in VR and technology has altered the structural components of the way the HMD interacts with the user [37]. Most VR HMDs, both for consoles and mobile devices, operate using a projection onto a small screen a few inches from the user's eyes. The new version of VR projects the imaging directly into the user's retina as opposed to on the screen in front of them [37]. These technological alterations are allowing individuals with visual impairments to experience virtual settings more clearly than their "real" life interactions. It is not naïve to think of VR as being able to change the lives of visually impaired individuals and the way people interact with the contemporary world. The impact VR will have on globalized culture in the very near future will not only drive technological understanding yet propel accessibility of information for all kinds of people.

2.14 Conclusion

The implementation of video games and VR learning in classrooms can aid in bridging the "digital divide". Video game-based and VR balance the interaction of enjoyment as well as intellectual development. By implementing restrictions, these platforms can help alleviate the disagreement between teachers, students, parents, and game developers regarding the content and instructional use of video games [6]. Many game genres, like first-person shooter games, in neither non-virtual nor virtual platforms, are not structured efficiently for learning due to their emotional interaction with violent content [44]. Yet, as previously discussed, benefits of video game playing can include improved attention and visual focus, increased understanding of executive decision making through strategic planning, and practice in collaborative contexts. [4,8]. Most commercial role-playing games are structured to improve attention and visual focus, with the structural emphasis on autonomy and self-regulated learning [4, 14, 16]. For some individuals, more autonomy within a game increases player satisfaction and lends to increased intrinsic motivation [14]. It has been shown, a guided path is beneficial for scaffolding learning and creating a balance between educational value and player satisfaction [14, 39–41]. However, it is the immediate statistical feedback that allows video games to be optimal as a learning environment. The combination of negative and positive evidence allows the learner to understand their mistake, but focus on game continuation [38].

The skills a player can acquire through video game and VR learning can include improved attention, increased executive decision making through strategic planning, self-regulation of learning, collaboration experience, and identity formation [4, 8, 34, 42, 45, 46]. As shown, the video game platform fosters motivation, effectiveness, and learner satisfaction while developing visual and digital learning skills [46]. These positive cognitive affordances are heightened in VR learning, allowing the learner to engage with the video game environment more critically. Ultimately, VR has the capability to improve relationships between populations and strengthening worldwide communication through language development if applied appropriately. More important, digital and visual literacies are developed through practice and interaction with these digital interfaces [2]. Utilizing these technologies primarily used in leisure activities within classroom education will help bridge the "digital divide" increasing in youth achievement and aide in the development of 21st century skill needed to thrive in today's world [3,5].

Families with more income can consistently buy innovative technologies and the newest video game consoles. Since not all households possess computers, internet, video games, let alone the newest VR console, a gap in digital interaction is established [5]. Children with access to video games at home will increase their visual and digital literacy skills, while others will not have this opportunity. This means families struggling financially cannot offer this type of learning experience to their youth, in turn, limiting their visual/digital literacies and learning potential [3]. Coincidentally, not all games are accessible to all types of learners with restrictions to access, languages used, and content [3].

When in the classroom, these differences in skills and digital literacy development are noted and usually interpreted as an intellectual or communicative deficit [43]. In other words, teachers are interpreting the lack of accessibility and development of 21st century skills outside of the classroom as struggles with general learning [3]. The skills gained from digital interaction and visual development will be necessary for thriving in the 21st century [3]. Therefore, implementing this type of learning and digital interaction in the classroom will aid closing the "digital divide" increasing in today's classrooms while providing an enjoyable experience for contemporary youth [5].

References

[1] Entertainment Software Association (2015). Essential acts about the computer and video game industry. http://www.theesa.com/wp-content/uploads/2015/04/ESA-Essential-Facts-2015.pdf Downloaded March 10, 2017.

[2] Gilster, P. and Glister, P. (1997). Digital literacy. New York: Wiley Computer Pub.

[3] Black, R. W. (2009). English language learners, fan communities, and 21st century skills. Journal of Adolescent & Adult Literacy, 52(8), 688–697.

[4] Belchior, P., Marsiske, M., Sisco, S. M., Yam, A., Bavelier, D., Ball, K. and Mann, W. C. (2013). Video game training to improve selective visual attention in older adults. Computers In Human Behavior, 29(4), 1318–1324.

[5] Tate, T. and Warschauer, M. (2017). The Digital Divide in Language and Literacy Education. In Language, Education and Technology (pp. 45–56). Springer, Cham.

[6] Zichermann, G. (2011). Gabe Zichermann: How games make kids smarter [Video File]. Retrieved from https://www.ted.com/talks/gabe_zichermann_how_games_make_kids_smarter

[7] Grandin, T. (2006). Thinking in pictures: And other reports from my life with autism. Vintage.

[8] Gee, J. P. (2017). Teaching, learning, literacy in our high-risk high-tech world: A framework for becoming human. Teachers College Press.

[9] Healey, C. and Enns, J. (2011). Attention and visual memory in visualization and computer graphics. IEEE Transactions on Visualization and Computer Graphics, 18(7), 1170–1188.

[10] McCloud, S. (2011). Making comics. Harper Collins.

[11] Rieber, L. P. (1996). Seriously considering play: Designing interactive learning environments based on the blending of microworlds, simulations, and games. Educational Technology Research and Development, 44(2), 43–58.

[12] Rosas, R., Nussbaum, M., Cumsille, P., Marianov, V., Correa, M., Flores, P. and Salinas, M. (2003). Beyond Nintendo: design and assessment of educational video games for first and second grade students. Computers & Education, 40(1), 71–94.

[13] Zap, N. and Code, J. (2009). Self-Regulated Learning in Video Game Environments, in Ferdig R.E. (Ed.) Handbook of Research on Effective Electronic Gaming in Education, pp. 738–756. Hershey, PA: IGI Global. Google Scholar.

[14] Chik, A. (2014). Digital gaming and language learning: Autonomy and community. Language Learning & Technology, 18(2), 85–100. Retrieved from http://llt.msu.edu/issues/june2014/chik.pdf

[15] Granic, I., Lobel, A. and Engels, R. C. (2014). The benefits of playing video games. American Psychologist, 69(1), 66.

[16] Vogel, J. J., Vogel, D. S., Cannon-Bowers, J., Bowers, C. A., Muse, K. and Wright, M. (2006). Computer gaming and interactive simulations for learning: A meta-analysis. Journal of Educational Computing Research, 34(3), 229–243.

[17] Singer, D. G., Golinkoff, R. M. and Hirsh-Pasek, K. (2006). Play = Learning: How play motivates and enhances children's cognitive and social-emotional growth. Oxford University Press.

[18] Engeström, Y., Miettinen, R. and Punamäki, R. L. (1999). Perspectives on activity theory. Cambridge University Press.

[19] Bergen, D. (2002). The role of pretend play in children's cognitive development. Early Childhood Research & Practice, 4(1), n1.

[20] Lillard, A. S. (1993). Pretend Play Skills and the Child's Theory of Mind. Society for Research in Child Development, 4(2), 348–371.
[21] Lave, J. and Wenger, E. (1991). Situated learning: Legitimate peripheral participation. Cambridge University Press.
[22] Decortis, F. and Rizzo, A. (2002). New active tools for supporting narrative structures. Personal and Ubiquitous Computing, 6(5–6), 416–429.
[23] Harris, P. and Daley, J. (2008). Exploring the contribution of play to social capital in institutional adult learning settings. Australian Journal of Adult Learning, 48(1), 50.
[24] McDermott, A. F., Bavelier, D. and Green, C. S. (2014). Memory abilities in action video game players. Computers In Human Behavior, 3469–3478.
[25] Green, S. C., Sugarman, M. A., Medford, K., Klobusicky, E. and Bavelier, D. D. (2012). The effect of action video game experience on task-switching. Computers In Human Behavior, 28(3), 984–994.
[26] Jackson, L. A., Witt, E. A., Games, A. I., Fitzgerald, H. E., von Eye, A. and Zhao, Y. (2012). Information technology use and creativity: Findings from the Children and Technology Project. Computers in Human Behavior, 28, 370–376. doi:10.1016/j.chb.2011.10.006
[27] Bissell, T. (2011). Extra lives: Why video games matter. Vintage.
[28] Gee, J. P. (2003). What video games have to teach us about learning and literacy. Computers in Entertainment (CIE), 1(1), 20–20.
[29] McGonigal, J. (2011). Reality is broken: Why games make us better and how they can change the world. Penguin.
[30] Squire, K. (2011). Video Games and Learning: Teaching and Participatory Culture in the Digital Age. Technology, Education – Connections (the TEC Series). Teachers College Press. 1234 Amsterdam Avenue, New York, NY 10027.
[31] Mitchell, A. and Savill-Smith, C. (2004). The use of computer and video games for learning: A review of the literature.
[32] Zimmerman, B. J. (2002). Becoming a self-regulated learner: An overview. Theory Into Practice, 41(2), 64–70.
[33] Zimmerman, B. J. and Schunk, D. H. (Eds.). (2001). Self-regulated learning and academic achievement: Theoretical perspectives. Routledge.
[34] Anderson, N. J. (2002). The Role of Metacognition in Second Language Teaching and Learning. ERIC Digest.

[35] Buckley, K. E. and Anderson, C. A. (2006). A theoretical model of the effects and consequences of playing video games. Playing Video Games: Motives, Responses, and Consequences, 363–378.

[36] Calvert, S. L., Appelbaum, M., Dodge, K. A., Graham, S., Nagayama Hall, G. C., Hamby, S., . . . and Hedges, L. V. (2017). The American Psychological Association Task Force assessment of violent video games: Science in the service of public interest. American Psychologist, 72(2), 126.

[37] Durbin, Joe. "Intel's New AR Prototype Uses Retinal Projection and Looks Cool Doing It." VRScout, 6 Feb. 2018, vrscout.com/news/intel-ar-prototype-retinal-projection/.

[38] Aljaafreh, A. and Lantolf, J. P. (1994). Negative feedback as regulation and second language learning in the zone of proximal development. The Modern Language Journal, 78(4), 465–483.

[39] Atkins, B. (2003). More than a game: The computer game as a fictional form. Manchester, UK: Manchester University Press.

[40] Newman, J. (2002). The myth of the ergodic videogame. Game studies, 2(1), 1–17.

[41] Neville, D. O. (2010). Structuring Narrative in 3D Digital Game-Based Learning Environments to Support Second Language Acquisition. Foreign Language Annals, 43(3), 446–469.

[42] Thorne, S. L., Sauro, S. and Smith, B. (2015). Technologies, identities, and expressive activity. Annual Review of Applied Linguistics, 35, 215–233.

[43] Philips, S. U. (2001). Participant structures and communicative competence: Warm Springs children in community and classroom. na.

[44] Morris, S. (2002). First person shooters – a game apparatus. In G. King & T. Krzywinska (Eds.), Screen-Play: Cinema/videogames/interfacings (pp. 82–85). London: Wallflower.

[45] Zimmerman, B. J. (1990). Self-Regulated Learning and Academic Achievement: An Overview, Educational Psychologist, 25(1), 3–17, DOI: 10.1207/s15326985ep2501_2

[46] Gee, J. P. (2006). Are video games good for learning?. Nordic Journal of Digital Literacy, 1(3), 172–183.

3

Virtual Reality and Movement Disorders

Rachneet Kaur[1,*], Manuel E. Hernandez[2] and Richard Sowers[3]

[1]Department of Industrial and Enterprise Systems Engineering,
University of Illinois at Urbana-Champaign, USA
[2]Department of Kinesiology and Community Health,
University of Illinois at Urbana-Champaign, USA
[3]Department of Industrial and Enterprise Systems Engineering,
Department of Mathematics, University of Illinois at
Urbana-Champaign, USA
E-mail: kaurrachneet6@gmail.com
*Corresponding Author

Falls are one of the prominent reasons inducing accidental fatality and fractures among the elder population and those with movement disorders. The proposed work is to study and attempt to alleviate the fear of falling in humans using experimental setups designed in VR. This chapter is primarily aimed at establishing and understanding the underlying fluctuations relative to anxiety in human postural control using immersive virtual conditions. We review the utilization of VR for evaluating the effects of fear of falling on balance and gait function. Further, we examine the potential for VR neurorehabilitation aimed at ameliorating fall-related anxiety in adults. We majorly discuss this in the chapter through analyzing real-time neurological feedback, measured via electroencephalogram (EEG) signals, in two experimental designs devised at studying neural responses relative to human postural threats, namely, visual cliffs while walking in a virtual world and induced randomized height changes and perturbations in a quiet standing environment. This framework incorporating virtual environments for understanding postural challenges is a step forward towards a smart and intuitive brain-computer interface (BCI) system that detects and dynamically adapts to the environment analogous to the real-time anxiety state of humans. Furthermore, through the integration of smart and intuitive BCI systems, potential breakthroughs in both physical and psychological recovery after falls may be feasible through greater access to personalized medical treatment and development of novel neurorehabilitation therapies aimed at treating excessive fear of falling, and hence, a potential source of fall risks, especially among older adults and those suffering with movement disorders, such as Parkinson's disease.

3.1 Introduction

This chapter seeks to understand individual cognitive and neural process variations in response to realistic and complex environments and sensorimotor integration alterations under anxiety-inducing conditions via VR-based frameworks.

3.1.1 Fear of Falling

As 10,000 Americans are turning 65 each day [1] and nearly one in every five U.S residents is projected to be aged 65 and older in 2030 [2], promoting well-being and healthy aging in older adults is becoming increasingly crucial to advance towards a better quality of life. Mobility and in particular fear of falling (FOF) remains one of the significant issues while trying to progress towards healthy aging. Several researches have explored interactions between mobility and healthy aging in the past [3–5]. A study in [6] determined that aerobic training facilitates clinical and physical improvements in depressive elderly adults, and hence protecting against a decrease in cortical activity. Experiments conducted by [7] on 21 elderly subjects aging 67 to 92 years observed that the margins to the spatial-temporal boundaries of postural stability decrease with advancing age, which may contribute to the progressive instability of posture with aging in the elderly. Experiments on 30 older adult subjects in [8] explored difficulty disengaging from fall-threatening stimuli among fall-fearful older adults. FOF, known to reduce balance performance in older adults, is a prominent risk factor for fall risks; which is a leading cause for injuries and mortality among the elderly [9]. It is often accompanied by physical and psychological effects that enhance the fall risk. Given this strong association between the FOF and fall risks [10], description of the mechanisms that govern anxiety-related changes in postural control is critical. Understanding these anxiety-related alterations may lay the foundations for the development of novel therapeutic procedures to help reduce fall risks, especially among older adults with movement disorders such as Parkinson's disease. FOF might also alter postural control strategies [11], studying changes in anxiety levels when at different heights and during induced perturbations might further help explain this correlation. Visual stimuli is known to cause anxiety in patients suffering from acrophobia or pathological phobia of heights, while they try to manage standing balance [12]. Acrophobia is one of the most prevalent and severe specific phobias affecting nearly 4.9% of the population [13]. Hence, studying anxiety-related responses to immersive visual stimuli and induced height and perturbation changes will aid in further understanding acrophobic responses. Research on Parkinson's disease patients have deduced that exposure to elevated heights inclines them to possibly fall in a backward or medial-lateral direction [14]. Therefore, understanding responses to height variations is also significant in designing remedial treatments for elderly and patients suffering from Parkinson's disease and other movement disorders [15].

3.1.2 Virtual Reality

Virtual reality (VR) is an experience where a person interacts with a controlled or modifiable artificial reality environment emulating the real world, where the subject responses can be monitored and evaluated [16]. Further, VR frameworks assist in isolating subjects from non-natural or complex laboratory settings [17, 18]. Hence, immersive VR environments, being secure and practicable to provide representative visual scenarios, have been investigated broadly by researchers to examine acrophobia and neural training [19, 20]. For people suffering from motor and mental health dysfunctions, VR training is potentially viable therapy. The underlying idea of VR-based therapeutic treatments for motor and cognitive disorders is to engage subjects in multisensory training, and hence aid to enhance neuroplasticity of the human brain. A comprehensive review of VR as a platform for neuromodulation and neuroimaging was presented by [21]. To demonstrate that VR can induce psychological responses analogous to real-world height controls for studying fall-related anxiety in older adults, a system that records neural, physiological, and behavioral data in an engaging virtual environment, while implementing sensory and mechanical perturbations corresponding to high postural threat conditions in the real world, was introduced in [22]. A realistic ocular feedback of cliff scenarios in VR has been studied for analyzing acrophobic responses of the human brain [23]. To understand dynamic neurological responses to visual cliffs while walking for mitigating FOF, especially among the elder population and those suffering with movement disorders, an experimental setup that monitors a subject's real-time neural activity via electroencephalogram (EEG) signals while being immersed in a virtual world and walking on an instrumented treadmill, was investigated by [24]. Results in [25] on 10 healthy young adults indicated an influence of psychological factors related to postural threat on the cortical activity associated with postural reactions to unpredictable perturbations. Further, a study conducted by [26] on 10 healthy individuals found insights into links between cortical and cognitive influences on compensatory balance control. Fall risk and postural stability experiments in [27] concluded the efficacy of a short-term VR-based balance training program on the balance ability of patients suffering from multiple sclerosis. The study concluded that the fall risk index and overall stability index of the patients significantly improved after 24 sessions of the VR balance training.

3.1.3 Brain-Computer Interfaces

To investigate anxiety responses and neurological feedback, some researchers have previously examined experimental setups unifying VR and EEG. Consequences of high heights exposure during beam-walking in a VR environment on physiological stress and cognitive loading was explored using statistical analysis on EEG signals in [28]. In experiments conducted by [29], a combination of VR with robotics-based rehabilitation induced an improvement in gait and balance among patients suffering from chronic hemiparesis. Statistical analysis on the EEG data collected during the experiments suggested that the use of VR may entrain brain areas responsible for motor planning and learning, and hence may potentially lead to an enhanced motor performance in humans. Moreover, EEG and VR-based brain-computer interface (BCI) systems, that allow brain responses to control virtual robots or surroundings, may serve in further facilitating neural rehabilitation. The work proposed in [30] concluded that a closed-loop EEG-based BCI-VR system enhances cortical involvement in human treadmill walking, triggers cortical networks involved in motor learning, and further enhances voluntary control of human gait. This study indicates that EEG has the capacity to monitor cortical activity in treadmill walking, hence enabling EEG-based BCI systems for walking as a paradigm for improving the rehabilitation adequacy. A few other VR inspired walking based BCI systems have been examined in the past for neurorehabilitation [30–32].

This chapter chiefly discusses two reliable and practicable experimental setups designed using VR essentially focused toward interpreting and demonstrating the effects of anxiety variations in postural control. The two novel experimental setups, primarily configured to inspect neural responses in postural threat conditions, specifically, the VR *walking experiment*, where subjects experience visual cliffs with varying depths while walking in a VR environment and the VR *height control experiment*, where subjects undergo induced randomized height changes and perturbations in a quiet standing virtual environment were initially proposed in [24] and [33] respectively. The VR *walking experiment* is a test setup dynamically modifying the virtual world using the subject's estimated neural responses (captured via online EEG measurements) to visual stimuli. The VR *height control experiment* is a VR-based test setup for studying subject's anxiety state, estimated via the recorded EEG signals, to randomized height changes (induced in VR) and mechanical perturbations (induced via tilting force plates) while they stand. These designed setups aid in better understanding some neural aspects of FOF via virtual environments resembling the real-world testing scenarios. In these experimental conditions, the subject's anxiety states were evaluated using the *frontal alpha asymmetry index* (FAA) of the EEG recordings. Comprehensive experimental validation of these setups may assist in administering VR BCI-systems to gait therapies and cures, especially for the elderly and those experiencing movement ailments.

3.2 Goals and Contributions

The major objective of the outlined experimentation is to design strategies that assist in alleviating the fear of heights and falling in humans. These systems may facilitate directing the use of VR-based BCI-procedures for clinicians and practitioners to prepare strategic gait therapies for motor disorders affecting human movement. These improvised clinical designs may hopefully aid in substantially reducing therapy related costs for the movement disorders in the future.

In this chapter, the experimental setups and major findings from the VR *walking experiment* and the VR *height control experiment* for understanding FOF (via the EEG responses) while walking and height fluctuations, originally introduced in [24] and [33] respectively, are reviewed. The remainder of the chapter is organized as follows. In Section 3.2, the computational infrastructure for the two reviewed experiments is broadly discussed. In Section 3.3, the design and data collection methodology of the setups for the walking (3.4.1) and the height control experiment (3.4.2) is elaborated. In Section 3.4, EEG data processing performed for both the experimental designs is explained. In Section 3.5, major findings and some applications of the proposed designs are discussed. In Section 3.6, some applications of the reviewed experimental setups and corresponding effects on the therapy costs are discussed. In Section 3.7, some of the challenges, open-ended questions related to the testbeds and proposed methods to progress towards the same are considered. Finally, in Section 3.8, concluding remarks and several future directions for the work are highlighted.

3.3 Computational Infrastructure

For the experimental designs, an HTC Vive VR headset [34], a 64 Channel EEG cap [35] and a wireless heart rate monitor [36] were adopted to render an immersive virtual world and simultaneously measure the electrical activity from the brain and the heart respectively. The VR environment was designed using Unity [37], Vizard [38] and Blender [39] software. For delivering high resolution and flawless immersive environments for the VR system, the Nvidia GEFORCE GTX 1070 graphics card [40] was used. Moreover, a C-MILL treadmill [41] was utilized for the VR *walking experiment* to let subjects walk at a self-paced speed while being immersed in the VR environment while a NeuroCom Clinical Research System [42] is used in the VR *height control experiment* to induce pseudo-randomized perturbations while subjects stood on a pair of force plates. Two distinct operating systems, connected over a local network, with one rendering the VR environment for the HTC Vive headset and another capturing the EEG signals for data analysis, were employed. In addition, one computer connected to the Neurocom was utilized for the VR height control experiment for inducing perturbations during a subset of trials to the subjects.

For illustration, Figure 3.1 depicts the data flow across different systems for the VR *walking experiment*. Treadmill communication was established over a TCP network. The coordinates of the virtual world were set up corresponding to the coordinates of the real world and the HTC Vive controllers' position was calibrated with the positions of the treadmill's safety rails. Moreover, the wireless heart-rate monitor was connected using multi-threaded C# and signal processing pipelines for the EEG recordings, to analyze the real-time anxiety states (Section 3.4.1), were implemented using Python 3.6 [43] with the support of several open-source libraries (MNE [44], scikit-learn [45], pandas [46] etc.).

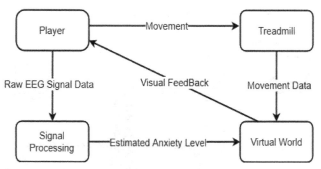

Figure 3.1 Data flow across different systems for the VR walking experiment.

3.4 Experimental Setups

In this section, the experimental setups for the VR walking and height control tests in subsections 3.4.1 and 3.4.2 was first suggested in [24] and [33] respectively, are summarized.

3.4.1 Visual Cliffs While Walking Experiment

The major goal of the VR walking experiment is to dynamically understand anxiety responses rendered by the human brain to height changes while walking in a virtual world via EEG biofeedback analysis. This research testbed guides the design of a self-regulated VR environment based on the subject's real-time anxiety level and allows for the monitoring and potential alleviation of anxiety during walking situations that necessitate balance and stability, especially for the elderly with high risks of FOF. The procedures for the experiment were approved by the University of Illinois Institutional Review Board. The proposed VR and EEG-based BCI framework consist of the following:

(1) An **HTC Vive VR headset** to present the virtual world.
(2) A **MotekForceLink C-MILL instrumented treadmill** in the self-paced mode to let subjects walk at an adaptive pace during the experiment. Also, the treadmill acquires online gait data for further analysis on subject's motion during the trials.
(3) A **Brainvision ActiCHamp EEG System** to record the EEG signals from 64 distinct sections of the brain.
(4) A **Zephyr HxM BT wireless heart rate monitor** to concurrently note the subject's heart rates.

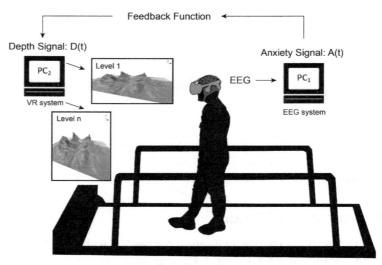

Figure 3.2   The brain-computer interface setup for the VR walking experiment. All subjects walk at a comfortable pace on a treadmill adaptive to their speed, while being immersed in a virtual world. The cliffs in the terrain modify depths to increase or decrease the scariness in response to the subject's real-time neural responses.

The BCI setup for the experiment is further illustrated in Figure 3.2. The designed environment is a terrain with a thin pathway, elapsing over various cliffs and mountains with varying depths and heights respectively. An example view of the terrain is shown in Figure 3.3. Corresponding to the online anxiety responses of the subjects, the terrain adapts to increase or decrease the scariness of the virtual world by adjusting the depths of the cliffs in transit. The higher anxiety level of the subject dynamically decreases the depth of pits and valleys in the virtual world, whereas, a comfortable pace hikes the depth of the cliffs in the terrain to appear frightening. A sample cliff along the pathway is shown in Figures 3.4 and 3.5. In the environment, a levelled terrain was used as a *neutral* pathway for baseline comparisons.

In the constructed terrains, the local minimum of the cliffs is adjusted along cubic Bézier splines [47, 48] (defined below).

Bézier curve of degree 1: A Bézier curve of degree 1 is defined by the linear interpolation between two points $\vec{P_0}$ and $\vec{P_1}$ defined in $\mathbb{R}^3$ (Equation 3.1).

$$B^{(1)}_{\vec{P_0},\vec{P_1}}(t) \overset{def}{=} (1-t)\vec{P_0} + t\vec{P_1} \tag{3.1}$$

Figure 3.3 **Left**: Setup with a subject walking the treadmill wearing the VR headset and EEG cap. **Right**: Virtual world.

Figure 3.4 Sample terrain: Overview.

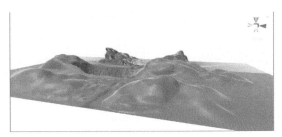

Figure 3.5 Sample terrain: Subject view.

Bézier curve of degree *n*: A Bézier curve of degree n is defined recursively by the linear interpolation between $B^{(n-1)}_{\vec{P}_0,...,\vec{P}_{n-1}}$ and $B^{(n-1)}_{\vec{P}_1,...,\vec{P}_n}$ (Equation 3.2) with n points $\vec{P}_0, \vec{P}_1, \ldots, \vec{P}_n$ defined in $\mathbb{R}^3$.

$$B^{(n)}_{\vec{P}_0,...,\vec{P}_n}(t) \overset{def}{=} (1-t)B^{(n-1)}_{\vec{P}_0,...,\vec{P}_{n-1}}(t) + tB^{(n-1)}_{\vec{P}_1,...,\vec{P}_n}(t) \qquad (3.2)$$

A sequence of cubic Bézier curves joined end-to-end is referred to as *cubic Bézier spline*. For further illustration, the path of deformation of the center of the area which needs to be distorted is depicted in Figure 3.6.

3.4.2 Height Control Experiment

For effectively understanding neural responses of the human brain while standing still to pseudo-randomized variations in height and depth conditions and induced perturbations, a VR-based testing setup with an EEG data collection system was arranged with the following infrastructure:

(1) A **safety harness** suspended to subjects for additional security during the experiment.
(2) An **HTC Vive VR headset** rendering the virtual environment with seven distinctive height conditions.
(3) A **NeuroCom SMART Equitest Clinical Research System** (Figure 3.7) to induce 10-degree toe-down perturbation during the trials.
(4) A **Brainvision 64 Channel ActiCHamp EEG system** mounted on the subject's head to record the electrical activity in the brain.

The data collection setup for the experiment is further illustrated in Figure 3.8. The protocols for the setup were approved by the Institutional

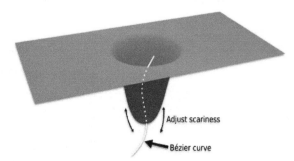

Figure 3.6   Deformation via cubic Bezier spline. This figure is adapted from [24].

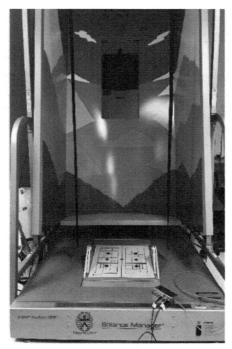

Figure 3.7 The NeuroCom SMART Equitest Clinical Research System. This figure is adapted from [33].

Review Board (approval number 17010). All subjects stood still on the NeuroCom force plate and experienced a podium in the virtual world (through the HTC Vive headset) aligned with the force place they were bodily standing on. Before the start of the actual experiment, baseline EEG data was recorded with subjects standing still and relaxing with eyes open for 90 seconds. Next, for each trial of the experiment, the virtual podium rose to 2.5 meters, 5 meters, or 7.5 meters for height conditions and the pit around the podium dropped to 2.5 meters, 5 meters, or 7.5 meters during the depth conditions. Hence, in total, eight height conditions including the ground level twice were experienced by all subjects in a single trial. Figure 3.9 illustrates the environment for two such conditions, namely, the ground level (left) and 7.5-meter height (right). Furthermore, to prevent the sequence of appearance of VR height and depth conditions biasing the empirical results, they were presented to the subjects in an anonymous pseudorandomized arrangement. Moreover, a 10-degree toe-down perturbation (as shown in Figure 3.10) was

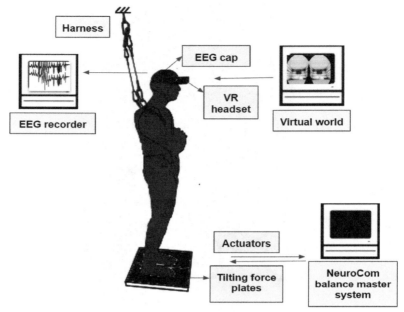

Figure 3.8   EEG data collection setup for the VR height control experiment. All subjects experience distinct height and depth conditions in a virtual environment in a pseudo-randomized order while standing on a force plate. Also, 10-degree toe-down perturbation was induced during half of the trials. This figure is adapted from [33].

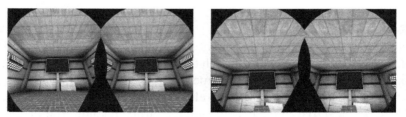

Figure 3.9   VR environment with podium at ground level (left) and at 7.5-meter height (right). This figure is adapted from [33].

induced at each condition during two of the four such trials that were run, where the subjects were instructed to sustain their stance while standing still.

The above-discussed settings were used for data acquisition and testing of the VR walking and height control experimental designs. Online EEG data from 64 distinctive sections of the brain were recorded for all subjects during

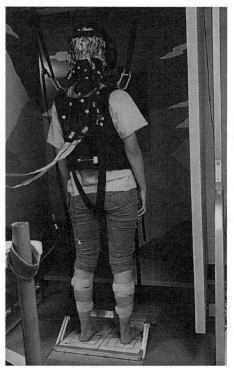

Figure 3.10 Subject experiencing a 10-degree toe-down perturbation via the NeuroCom force plate during the VR height control experiment.

each of the two setups. Next, each subject's induced anxiety was analyzed using the EEG analysis pipeline, as discussed further in Section 3.4.1.

3.5 Data Analysis

In this section, the EEG and heartbeat data analysis pipelines to evaluate the fluctuations in anxiety levels during both the described experimental setups are discussed.

3.5.1 EEG Processing

Using a 64 electrode EEG head cap, with channels as depicted in Figure 3.11, EEG signals were recorded from 64 discrete regions of the subject's brain. The recorded EEG signals at a 1000 Hz frequency were

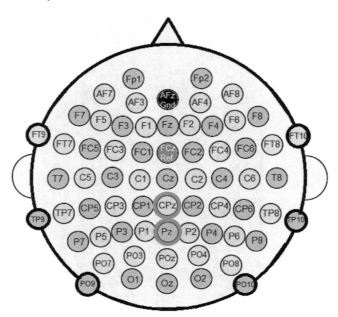

Figure 3.11   Electrodes in the 64 channel EEG head cap (refer [49]). Here, *green color* denotes electrode positions for channels 1 to 32, *yellow color* for channels 33 to 64, *blue* electrode depicts the reference channel and *black* depicts the ground electrode. Also, frontal channels F3 and F4 are representative of the *frontal alpha asymmetry index* (FAA).

first processed to time sync with the actual experiment and eliminate the peripheral signals. The retained 64-dimensional signals $\{X^{(n)}\}_{n=1}^{64}$ are refined via a finite impulse response filter (FIR) [50]. Next, Independent Component Analysis (ICA) [51] is employed to exclude the artifacts or independent components reflecting non-brain segments [52–55] in the obtained filtered signals $\{Z^{(n)}\}_{n=1}^{64}$. The preserved brain components are back-projected to extract the *frontal alpha asymmetry index*, representative of anxiety in humans [56, 57]. A detailed EEG analysis pipeline is further illustrated in Figure 3.12. The major procedures adopted in the EEG analysis pipeline are further explained below.

Signal preprocessing: Due to the time lag introduced in the start of the experiment for providing instructions to the subjects, the recorded EEG signals $\{X^{(n)}\}_{n=1}^{N}$ are first synced with the start time of VR software and the actual experiment. Next, additional peripheral signal recordings from the auxiliary channels are eliminated to refine the EEG.

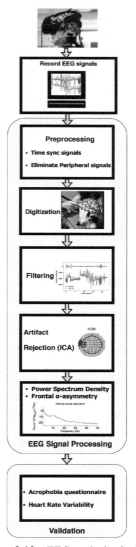

Figure 3.12 EEG analysis pipeline.

Signal Filtering: Next, the recorded noisy EEG signals are filtered using Fast Fourier transforms within the desired range in a feed-forward approach (Equation 3.3). Define

$$X_t^{(n),f} \overset{\text{def}}{=} \left(X^{(n)} * B \right)(t) \tag{3.3}$$

where B has Fourier transform corresponding to the desired bandpass filter. In particular, a bandpass finite impulse response filter (FIR) within 1 Hz to 45 Hz with a sampling frequency of 1 kHz, is used to refine the EEG signals.

Finite impulse response: At each value j, output of an Nth order FIR filter $x[j]$ (Equation 3.4) is a convolution of the $N + 1$ most recent input signal values $a[j - m], 0 \leq m \leq N$ using the impulse response values c_m as the filter coefficients at each instant $0 \leq m \leq N$.

$$x[j] = \sum_{m=0}^{N} c_m a[j - m] \tag{3.4}$$

FIR filters are a suggested preference in literature for the EEG signal filtering process [58].

Artifact Rejection: Given the potential for muscle and movement artifacts to introduce large sources of electrical noise in EEG recordings, it is imperative to carry out artifact removal techniques in EEG analysis [59]. Typical techniques are primarily focused on reduction, rejection, and recovery [60]. EEG data, being a statistically linear combination of cortical signals [61], independent component analysis (ICA) is executed on the filtered EEG data to extract 64 sub-components maximizing statistical independence (via *entropy* methods) and reject ocular, cardiac, muscle or extrinsic artifacts [52]. In addition to ICA, artifacts are further rejected using the variance, kurtosis and skewness calculations. Hence, only independent components (ICs) picking up signals relative to brain segments are preserved for further analysis to extract human anxiety state.

Independent Component Analysis: Let $[X^{(n),f}]_{64 \times T}$ be a matrix of recorded EEG data, where columns correspond to different channels and rows correspond to time. ICA is utilized to write the $X^{(n),f}$'s as a superposition of N independent signals. Formally, ICA seeks to write:

$$X^f = W^{-1}Y$$

for an invertible $N \times N$ matrix W^{-1}, known as mixing matrix, where

$$X^f = \begin{pmatrix} X^{(1),f} \\ X^{(2),f} \\ \vdots \\ X^{(N),f} \end{pmatrix} \quad \text{and} \quad Y = \begin{pmatrix} Y^{(1)} \\ Y^{(2)} \\ \vdots \\ Y^{(N)} \end{pmatrix}$$

where the $Y^{(n)}$,'s are statistically independent ICs reflecting different sources of the neural activity of interest and other artifactual activities as eye movements, neck muscle activity, sensor noise, and movement-induced artifacts. The separating weights matrix W is computed such that the statistical independence between the components of

$$Y = WX^f$$

is maximized, measured by minimizing an entropy with respect to i.i.d. uniform random variables, overall invertible matrices W.

To retain only the ICs corresponding to brain responses for further analysis and remove any artifactual components relative to eye, cardiac or muscular activities, $Y^{(\tilde{n}^*)}$, the "bad" ICs relative to artifacts are detected (via the *IC representation diagram*). Next, define Π to be the diagonal projection matrix with

$$\Pi_{n,n} = \begin{cases} 1 & \text{if } n \neq \tilde{n}^* \\ 0 & \text{if } n = \tilde{n}^* \end{cases} \tag{3.5}$$

Then, $Y_0 = \Pi Y$ agrees with Y except that the $\tilde{n}^*$-th component is removed. Hence

$$X_0^f = W^{-1}Y_0 = W^{-1}\Pi Y = W^{-1}\Pi W X^f \tag{3.6}$$

corresponds to reconstructed EEG signals after removing the effect of the reject-able ICs. The artifact-free remixed signals X_0^f are used to compute the power spectrum density.

An *IC representation diagram* supports its classification into an acceptable or a rejectable IC. It consists of the following elements (refer [62] for a more descriptive explanation of components of an IC description image).

(1) A **scalp topography** heat map demonstrating the response of ICA on each of the EEG electrodes. The columns of mixing matrix W^{-1} are used to describe the scalp projections. In the diagram (say the top left image in Figure 3.13), black dots signify the positions of respective electrodes with red, blue and white colors denoting positive, negative and null contributions respectively.

(2) An **event-related potential (ERP)** image reflecting the IC's activity over the complete data set.

(3) A **time series** depicting a segment of the IC's activity, somewhat similar to an ERP image.

(4) An **activity power spectrum** plot displaying the power dispersion of the IC's activity with respect to frequencies over the complete data set.

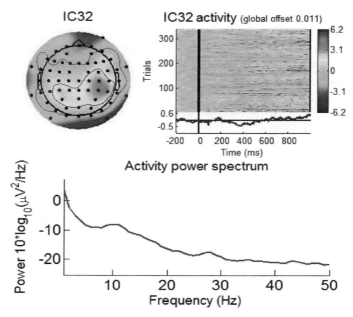

Figure 3.13   An accepted independent component. The IC representation consists of a scalp topography image (top left), an event-related potential (ERP) image (top right), a time series displaying activity from the component (middle right, below the ERP diagram) and an activity power spectrum plot (bottom). In the scalp projections diagram, red, blue and white colors denote positive, negative and null contributions respectively. The heat map demonstrates inputs from multiple electrodes. This figure is adapted from [24].

Figures 3.13–3.15 and 3.16 depict the topographical plots (top left), ERP image (top right), time series depicting the component activity (middle right, below the ERP diagram) and power spectrum (bottom) for four different independent components obtained after performing ICA on the EEG data.

The separated ICs can be classified in the following seven distinct groups corresponding to the source of the EEG activities (refer [63, 64] for a more detailed explanation on the distinction of ICs in the EEG data).

(1) The spatially consistent EEG signals originating from the cerebral cortex are classified as *brain components*. The scalp topography of such components depicts continuous and wide significant contributions from multiple channels of the brain [65]. Further, the power spectrum for these components is inversely proportional to the frequency and commonly observes a peak in the αband of frequency (8–12 Hz). Figure 3.13

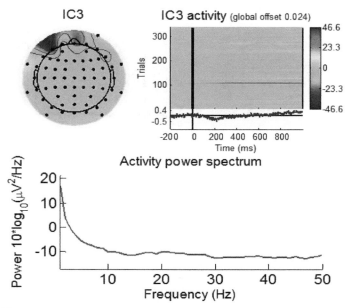

Figure 3.14   A rejected independent component representing a muscle artifact from the face of the subject.

illustrates an example of a *brain component* with broad contributions in the scalp topography from more of the channels and power spectrum plot depicting an increase at 10 Hz, followed by a consistent decline. Hence, it is labeled as a *brain component* and included in the computation of power spectrum density.

(2) The electric signals introduced from the retina, representing vertical or horizontal eye motion are grouped as *eye components*. The power spectrum for these components is usually observed at frequency values lower than 5 Hz. The time series of components relative to vertical and horizontal eye movements are likely to involve spikes signifying eye blinks and steps indicating visual scanning respectively.

(3) The EEG signals generated by muscle activities are classified as *muscle components*. The power spectrum for these components is usually observed at frequencies higher than 20 Hz. Figures 3.14 and 3.15 depict the examples of a *muscle component*. Figure 3.14 demonstrates a muscle artifact from the face of the subject, given the spatial distribution of weights, and a large magnitude of activity found in a single trial, which

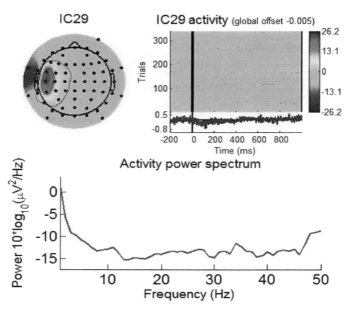

Figure 3.15 A rejected independent component representing a muscle artifact from the side of the head. The power spectrum plot reflects no apparent peaks in the α-band. This figure is adapted from [24].

leads to its rejection. Figure 3.15 shows narrow bounded activity in the scalp topography diagram, indicating a muscle artifact from the side of the head. Hence, it is excluded from further analysis of the signals.

(4) ICs signifying the electric signals produced from the heart are classified as *heart components*. The power spectrum for these components generally involves no peaks.

(5) ICs generated from external electrical fields such as electronics are termed as *line noise*. The power spectrum for these artifacts mostly involves peaks at 50–60 Hz frequency range.

(6) Artifacts introduced via impaired EEG head cap electrodes or detached electrode connections at the time of the experiment are referred to as *channel noise*. The scalp topography for such components concentrates most of the focal weight on a single EEG electrode. Figure 3.16 is an example of a *channel noise* component as it demonstrates noise from a single electrode, as indicated by focal weights in the IC and erratic electrical activity in 1–2 trials. The power spectrum plot has a significant increase of around 50 Hz, and no evident increases in activity in either

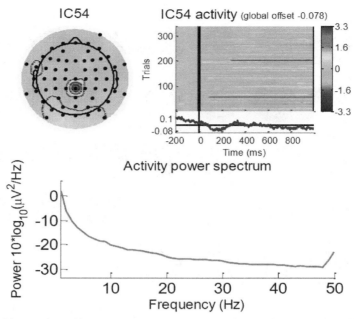

Figure 3.16 A rejected independent component representing noise or strong weight from a single electrode.

$\theta(4-8$ Hz$)$, $a(8-12$ Hz$)$, or $\beta(12-30$ Hz$)$ frequency ranges, which leads to its rejection.

(7) Any other type of ICs that do not confine to above-mentioned classes but represent a noisy signal are categorized as *other artifacts*.

Only the *brain components* are classified as "good" acceptable ICs that are back-projected to compute the power spectrum density and all the other classes are grouped as "bad" ICs that are rejected from any further analysis. Hence, Figure 3.13 represents an acceptable IC and Figures 3.14–3.16 are examples of rejected components. An overview of the first 32 independent components is shown in Figure 3.17. The components in green signify accepted ICs and the red ones are rejected artifacts.

In particular, one big challenge in EEG analysis is automatic IC detection. To date, research has been focused on automatic identification and removal of eye movement and blink artifacts from EEG signals [66], automatic and direct identification of blink components from scalp EEG [67], and automated EEG artifact elimination through the application of machine learning algorithms to ICA-based features [68]. Building upon blind source

Figure 3.17   An overview of the first 32 independent components from an EEG analysis pipeline. The components in *green* are accepted and the *red* ones are rejected artifacts. Only the accepted components relative to capturing brain activities are back-projected to compute the power spectrum density.

separation techniques, a new ICA-based fingerprint method for the automatic removal of physiological artifacts from EEG recordings has been recently introduced [69], while crowd labeling latent Dirichlet allocation, a generalization of latent Dirichlet allocation, has been applied to EEG IC labeling [64]. Further details regarding the same are discussed later in Section 3.7.1.

Frontal alpha asymmetry: Power Spectrum Density (PSD) has been adopted in the past to identify relationships with the drowsiness level, vigilance and sharing behavior via the EEG data [70, 71]. In this work, the PSD amplitudes in the α-frequency range (8–12 Hz) are used to measure the frontal alpha asymmetry (FAA), which is observed in the literature to be a pertinent indicator of anxiety, expressing withdrawal and approach in humans [56, 57, 72, 73].

Frontal alpha asymmetry: FAA is defined as the difference of natural log of the PSD amplitudes in the α-frequency range from channels on the left and right frontal cortex of the brain, namely electrodes F3 and F4 respectively

(Equation 3.7).

$$FAA = \ln\left(\sum_{i=8}^{12} P_i^{F3}\right) - \ln\left(\sum_{i=8}^{12} P_i^{F4}\right) \tag{3.7}$$

where P_i^C denotes the PSD amplitude of the channel C at the i-th frequency value.

3.5.2 Validation

In order to further validate the practicality of using FAA of the EEG data for detecting anxiety in humans while being immersed in the discussed VR-based experimental setups, recorded heart rate variability (HRV), a well-known biomarker for anxiety and physiologic stress [74] and a self-reported Acrophobia Questionnaire (AQ) [75] were utilized.

Note that the heart rate recordings during the experiment were normalized with the baseline values noted before the start of the experiment. Also, each subject completes an AQ consisting of 17 questions with numerical scores ranging from 0 to 6, which are then aggregated to quantify a self-evaluated anxiety state of the subject.

3.6 Experimental Results

In this section, we review the experimental results from the two discussed test setups, initially presented in [24] and [33] respectively.

3.6.1 Visual Cliffs While Walking Experiment

For validation of the designed VR-based walking test setup, EEG data from a healthy young adult (HYA) subject walking the treadmill with the VR headset on was recorded while being immersed in environments reflecting control stimuli and anxiety-inducing situations. Computed EEG-based FAA values for the control stimuli (average $= 0.0415 \pm 0.0326$) and anxiety-inducing condition (average $= 0.07948 \pm 0.1577$) conclude that variance fluctuations from a near-zero baseline are noticeably higher in the anxious VR environments than the control VR-based circumstances. Figures 3.18 and 3.19 represent the box plots depicting median and variations in FAA obtained during the control stimuli and anxiety inducing VR environments presented to the subject respectively. These plots further validate the hypothesis that the

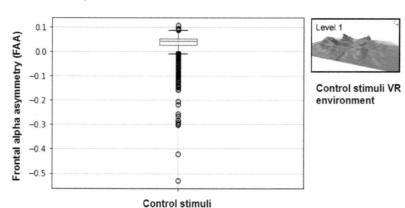

Figure 3.18   Box plot depicting variations in the FAA obtained while presenting the control stimuli to the subject.

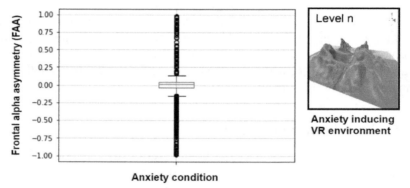

Figure 3.19   Box plot depicting variations in the FAA obtained while presenting the anxiety inducing VR environment to the subject.

FAA values, a potentially promising criterion for interpreting real-time anxiety state in humans, are consistent with the receptive landscapes presented to the subject in VR.

3.6.2 Height Control Experiment

For validating the VR height control experiment, EEG data (while immersed in the test environment) was collected via 5 HYA subjects with mean age $= 20.4 \pm 0.80$ years (40% females) and 1 male healthy old adult (HOA) volunteer aged 79 years from the regional neighborhood. Each subject's

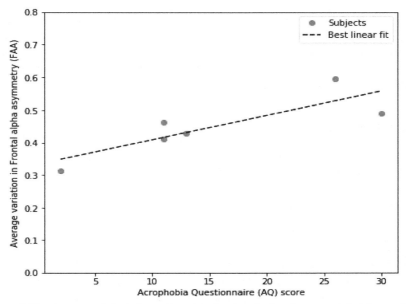

Figure 3.20 Average of deviations of anxiety levels calculated over each height condition with respect to the AQ scores. A linear fit with $R^2 = 0.71$ and $p = 0.04$ is suggested. This figure is adapted from [33].

real-time anxiety level was calculated via FAA with a moving window of 4 seconds succeeding ahead a second in each computation. As a result, a linear dependence with coefficient of determination $R^2 = 0.71$ and $p = 0.04$ between the average variations in FAA (calculated over each height condition over all trials) and self-reported AQ scores were noticed (as depicted in Figure 3.20), with a further stronger linear dependence with observed $R^2 = 0.79$ and $p = 0.01$ for trials with induced perturbations. This observed trend is in agreement with the hypothesis that effort and risk in maintaining balance is especially heightened during trials with induced perturbations [76]. Moreover, fitting an ANOVA [77–79] further verifies with $F = 7.78$ and $p = 0.04$ that the standard deviation in FAA experienced during trials with induced perturbations is correlated with the self-stated AQ scores by the subjects. Next, categorizing the subjects in three subgroups on the basis of their AQ scores, namely cohorts with *lower, medium and higher* AQ scores, it is observed that the subjects with a higher AQ experience a greater average deviation in FAA. Also, higher FAA deviation is observed (and validated via the paired t-test) when the subjects experience 7.5 meters height (*lower*: 0.353, *medium*: 0.468 and *higher*: 0.696) than when they experience 7.5 meters depth (*lower*: 0.259, *medium*: 0.415 and *higher*: 0.511). The above-observed effect may be attributed to neural disruptions caused via the contradictory indications between optically going up and accelerometry staying at the ground level [80, 81].

3.7 Discussion

In this section, effects on the therapy costs and some applications of the discussed VR-based experimental setups are explored.

3.7.1 Medical Care Costs

Fall-related medical treatment among older adults, and especially among older women, are associated with substantial economic costs [82–84]. In 2000, direct medical costs associated with fall-related injuries in the United States exceeded 19 billion dollars [82], with most costs associated with fractures [82–84]. Following a fall, older adults are more likely to suffer a significant injury such as a hip fracture or head injury [85, 86] that may require significant clinical intervention, surgery, extended hospital stay, and significant physical rehab [82]. However, after falling, anxiety about falling remains, which significantly impacts the quality of life of the individual [87, 88], particularly in persons with Parkinson's disease [15].

Through the integration of VR and BCI technologies, potential break-throughs in both physical and psychological recovery may be feasible through greater access to personalized medical treatment. Currently, VR technology has been successfully integrated with rehabilitation programs to provide improvements in both physical and cognitive functioning [89–92], as well as fear of falling [92–94]. In addition, symptoms arising from anxiety disorders such as acrophobia, arachnophobia, aviophobia, and social phobia have been ameliorated through the use of VR exposure therapies [95–110], and maybe more effective in some cases than relaxation therapy [111], although further evidence is needed [112–114]. Through the unification of BCI to VR environments, it has been demonstrated that kinesthetic motor imagery can be decoded to control the ambulation of an avatar within a VR environment [32] and maybe feasible in persons with spinal cord injury [31]. Furthermore, recent work has demonstrated the feasibility of using a closed-loop EEG-based BCIVR approach for controlling a walking avatar, promoting cortical engagement, and monitoring cortical activity while walking [30]. Given these observed benefits in BCI-VR, monitoring anxiety while standing or walking may provide a breakthrough in the treatment of physical and psychological symptoms in older adults.

Furthermore, the discussed experimental designs use head-mounted display (HMD) devices. HMD devices reduce the projected VR setup cost from millions of dollars to a few thousand [115]. Also, incorporating only the frontal channels for estimating the anxiety state rather than using all the electrodes on the EEG head cap considerably reduces cost, decreases burden on the subject and helps shorten the setup and computation time. Especially for the testbeds requiring motion (as the discussed VR *walking experiment*), wireless and "wearable" transmission systems for the EEG data collection are preferred and hence reducing the number of channels benefits such wireless systems [116] and may lead to novel applications in the "wearable" EEG technologies domain [117].

3.7.2 Clinical Implications of the Experimental Setups

Using the discussed combination of VR and actuated feedback via BCI based setups, the ultimate goal is to present an instantaneously adaptable scenario to the brain, focusing on the synthesis of real-time sensing and reaction. In particular, the considered frameworks have several broad objectives, listed as follows:

- Develop and validate electrophysical measures of anxiety in a realistic environment.
- Develop a geometric framework for understanding neural connectivity in response to anxiety.
- Develop a low-order feedback model that will frame quantitative thinking about learning and adaptation with respect to certain types of anxiety.

Moreover, the VR *walking experimental* design demonstrates methodologies for evaluating VR headsets in walking conditions, especially suited for balance training [28]. BCI setups, in general, have several inherent problems, namely the placement of the electrodes is imperfect, and the electrodes sometimes suffer from loss of connectivity. These can be problematic if they interfere with the rejection of unwanted artifacts. Some of these challenges and unsolved problems are further discussed in the next section.

3.8 Challenges, Open-Questions and Proposals

The discussed VR-based framework outlines a number of novel challenges in applied data analysis. For instance, a tool of dimensional reduction that identifies robust features of extreme events and further a notion of coarse-grained connectivity in mobile networks and how it is affected by extreme events needs to be understood. Some open-ended issues in the framework and proposed methods to make advancements towards the same are highlighted in this section.

3.8.1 Procedural Treatment of Artifacts

The EEG analysis pipeline allows us to sequentially remove different unwanted ICA components corresponding to artifacts, keeping in mind variations from individual to individual. This motivates defining a procedural way to formalize Equations (3.5) and (3.6) in removing artifacts. By having the subject walk in a visually non-threatening environment (no visual cliffs), neck muscle movements may be determined. After identifying and removing the neck artifacts, introducing visual cues (e.g., a bird) may identify signatures of eye movements. Attention to auditory movement could be identified by doing something like making the bird chirp. Another interesting question is the best sequence of these signal identification steps. Assuming that the neck muscle signatures are the strongest, that may be considered for removal first. To remove the next strongest signal, either Equations (3.5) and (3.6) could be recalculated using the same transformation matrix W, or infact W could be recalculated, in which case, N − 1 independent signals would be created from the N original signals. More generally, a sequence of unde-termined (sparse) ICA's may be calculated [118]. The challenge here is to establish the best sequence and an automated way to eliminate the unwanted ICA components represented in the EEG signals. Another potential idea to explore for automatic artifact rejection using the *IC representation diagram* is to apply supervised transfer learning [119, 120] based classification using convolution neural networks (CNN) [121] trained on the scalp topography plots of the ICs for the same experimental setup all across. This approach would require constructing an extensive IC image data set and providing the ground truth labels via experts labeling the IC properties image into distinct "good" brain and "bad" reject-able components, as categorized in Section IV-A.3.

3.8.2 Response of the Brain and Causality

In parallel to extracting independent components from the EEG signals; to better understand how signals propagate through the brain, observable signatures of neural connectivity need to be explored. A regression

$$X_{t+1}^f \approx \mathrm{M}X_t^f + \alpha D(t) + \xi_t \tag{3.8}$$

for $\mathrm{M} \in \mathbb{R}^{N \times N}$, $\alpha \in \mathbb{R}^N$, X^f as defined in Equation (3.3), D as the observed depth of the terrain and ξ as "noise" may be tried. Then $\mathrm{M}_{n',n}$ explains the sensitivity of n-th EEG signal to the prior n'-th. In certain regimes, where the sampling rate is much higher than the rate at which neural activity changes, Equation (3.8) might be replaced with

$$X_{t+1}^f \approx X_t^f + \mathrm{M}X_t^f \delta + \alpha D(t)\delta + \xi_t \sqrt{\delta}, \tag{3.9}$$

which leads to a stochastic differential equation

$$dX_t^f = \mathrm{M}X_t^f dt + \alpha D(t)dt + d\xi_t \tag{3.10}$$

where ξ_t denotes an appropriate stochastic process.

Models in Equations (3.8) and (3.10) may be combined with the ICA analysis of subsection IV-A.3 to understand dynamic response to anxiety (see [122]) and hence may help explain how anxiety propagates through the brain. By understanding the timescales of anxiety, the timescales at which the virtual depth should be changed to have different effects can be better deduced. Further, it would be interesting to study the interaction between the geometries of the relevant rows of W^{-1}, the matrix M, and the vector a [123].

3.8.3 Response of the Brain and Low-Rank Description

Low-rank factorization can give interesting insight into the complexity of matrices stemming from data [124]. Starting from Equation (3.8), a lower-rank description of the 64×64-matrix M may be studied. Using either Equations (3.8) or (3.9), M can be decomposed as

$$M = WH$$

or

$$M = I + WH\delta \tag{3.11}$$

where $W \in \mathbb{R}^{64 \times N'}$ and $H \in \mathbb{R}^{N' \times 64}$ for some $N' < N$; this ensures that M is of rank at most N'. Informally, W captures what parts of the brain are excited by neural connections, and H captures what information from the past is used in neural propagation. Mathematically, W and H can be computed by minimizing

$$E_\circ(W, H, \alpha) \stackrel{\text{def}}{=} \frac{1}{2T} \sum_{t=1}^{T} \|X_{t+1}^f - \{WHX_t^f + \alpha D(t)\}\|^2 \qquad (3.12)$$

overall $W \in \mathbb{R}^{64 \times N'}$, $H \in \mathbb{R}^{N' \times 64}$, and $\alpha \in \mathbb{R}^N$. Next, a penalty to Equation (3.12) might be added to drive some of the coefficients of H to 0 (i.e., to make H *sparse*); Lasso penalty [125] gives

$$\min_{\substack{W \in \mathbb{R}^{64 \times N} \\ H \in \mathbb{R}^{N' \times 64} \\ \alpha \in \mathbb{R}^N}} \left\{ \frac{1}{2T} \sum_{t=1}^{T} \|X_{t+1}^f - \{WHX_t^f + \alpha D(t)\}\|^2 + \lambda\|H\|_{L^1} \right\}$$

where $\|H\|_{L^1}$ is the sum of the absolute values of the entries of H. As $\lambda \gg 1$, the $L - 1$ penalty rewards sparsity of H; as a result, the dynamics of X_{t+1}^f involve less information from the prior state X_t^f. As λ is increased, M is approximated using fewer and fewer columns of W, and the actual error E_o increases. One measure of the complexity of neural interactions is the trade-off between N', λ, and E_o. In particular, for each value ϵ of E_o, a curve in (N', λ) space which gives ϵ may be studied for the complexity of regression in Equation (3.8).

3.8.4 Response of the Brain, Local Causality and Persistent Homology

Topography of the brain may also be considered in the above calculations. Matrix M is nonzero only if nodes n and n' are adjacent to each other, which enforces the requirement that interactions be local. The local interactions can be thought of as "paths" through the brain. Starting from local interactions, methods of *topological data analysis* can be used to identify robust global interactions [126]. $M_{n',n}$'s can be used to construct a connectivity graph that can be analyzed via the *barcode diagrams*. This might be a way to find robust structures in the presence of cortical noise. By fixing $\lambda > 0$, a collection of paths throughout the brain that follow nodes of sensitivity of magnitude at least λ may be constructed. Following a similar analysis as to [127], let $P(\lambda)$

consist of finite sequences $\gamma \overset{\text{def}}{=} (n_1, n_2 \cdots n_K)$ such that $|M_{nk, n_{k+1}}| \geq \lambda$. If $\gamma = (n_1, n_2 \cdots n_K)$, $-\gamma$ can be defined as

$$-\gamma \overset{\text{def}}{=} (n_K, n_{K-1} \cdots n_1)$$

and if $\gamma = (n_1, n_2 \cdots n_K)$ and $\gamma' = (n'_1, n'_2 \cdots n'_{K'})$ with $n_K = n'_1, \gamma_1 + \gamma_2$ can be defined as

$$\gamma_1 + \gamma_2 \overset{\text{def}}{=} (n_1, n_2 \cdots n_K, n'_2 \cdots n'_{K'})$$

If a path γ starts at some electrode position n, then $\gamma + (-(-\gamma)$ also ends an. For each ordered pair (n_{start}, n_{end}) of nodes, all paths which begin at n_{start} and end at n_{end} can be considered. Two such paths $\gamma1$ and $\gamma2$ are *equivalent* if $\gamma1 - \gamma2$ is homologous to the zero path through such paths of sensitivity at least λ. Informally, this would fail if $\gamma1 - \gamma2$ encloses an electrode which is connected by surrounding nodes with sensitivity less than λ. As $\lambda \nearrow \infty$, the connectivity decreases. A *persistence* barcode diagram records how this connectivity fluctuates with λ and captures the connectivity which is robust in λ. Prior work in connecting persistent homology to path structure for understanding the connectivity of *traffic congestion* via barcode diagrams is done in [128]. These ideas may lead to some useful ways to robustly capture the geometry of interactions between the signals at different electrodes. Robust ways of understanding connectivity could help reduce noise and might provide ways to overcome problems with sensor connectivity.

3.8.5 Feedback Signal

Given a way to measure a candidate signal for anxiety, a model to study feedback needs to be investigated. Letting A be the "anxiety" signal, a naïve model of the form

$$\dot{A}(t) = -\frac{1}{\tau}(A(t) - A_o) + \beta(D(t) - \underline{D})^+ \tag{3.13}$$

where

- τ is a relaxation time constant which captures a person's adaptability
- A_o is a steady-state anxiety level

- β is sensitivity to depth
- $\underline{D}$ is a reference acrophobic depth, may be explored. Visual cliffs that are shallower than $\underline{D}$ create no anxiety.

There are a number of aspects of this model that have natural interpretation. Informally, it may be assumed that older people take longer to adapt, meaning a larger τ. A larger value of $\underline{D}$ means that a person is less acrophobic. The baseline value A_o of A might be measured in the absence of any stimuli. D may be used to map out the stability boundaries of Equation (3.13), namely, setting

$$D(t) = \phi(A(t)) \tag{3.14}$$

for some appropriate function ϕ and considering the dynamics

$$\dot{A}(t) = -\frac{1}{\tau}(A(t) - A_o) + \beta(\phi(A(t)) - \underline{D})^+ \tag{3.15}$$

The structure of ϕ needs to be determined to most efficiently identify other parameters in Equation (3.13). A more realistic dynamics might include another dimension reflecting a "momentum". Within the proposed framework, it could be understood what aspects of neural signals can be affected by training and behavioral modification, and which might be intrinsic or "hard-wired".

3.8.6 Machine Learning

Given the recent progress of machine learning (ML) and deep learning algorithms in the medical domain for predicting various disorders [129, 130], these methods have the potential to analyze and classify real-time anxiety states. Previous works have explored ML methods to analyze features in EEG, for instance in stress detection [131] and hypnotic susceptibility classification [132]. Possible further work may include examination of these methodologies to validate EEG signals in classifying and predicting the onset and progression of real-time anxiety states [133], possibly via adopting heart rate variability (HRV), captured using the electrocardiography (EKG) and co-contraction index (CCI), extracted from the Electromyography (EMG) signals as the ground truth predictors for the neural responses in humans. While using the HRV as the ground truth with EEG data, separation of scales can be one challenge to further examine into, given the EEG data varying at a much faster scale than the recorded heartbeats.

3.9 Conclusions

In this chapter, we review and discuss the effectiveness of using VR-based BCI setups with FAA neurofeedback (computed via the EEG data) to help relieve fear of falling and heights in healthy adults while walking or standing still [72, 134, 135]. The described experimental and data analysis frameworks provide insights into neural signal geometry, neural connectivity, adaptation and noise in the neural signals. These setups in VR may guide clinicians towards developing novel acrophobia therapies and human postural control enhancement technologies. Further, these advancements may provide a breakthrough in the treatment of physical and psychological symptoms in older adults and those afflicted with Parkinson's disease and other movement disorders and impairments. Although, further analysis via accelerometry data is needed to establish if increased anxiety levels negatively affect postural control, if postural control impairment boosts anxiety, or if the two stimulate each other. These results would encourage clinicians to integrate BCI methods and neurofeedback to control anxiety and provide a treatment of acrophobia if the former holds, or practice postural and balance training to treat anxiety if the latter is valid. Further, using FAA as a marker for anxiety, and hence lesser number of electrodes in the head cap, helps reduce the setup related and computational hours. Future work involves verification of the examined test setups, to mitigate the fears relative to falls and heights, for elderly cohort and participants suffering from motion-related disorders.

Acknowledgments

We would like to thank all the subjects who collaborated with their time and data for the study, and JUMP ARCHES for financial support. This work would not have been possible without the support of the Illinois Geometry Laboratory of the Department of Mathematics at the University of Illinois.

References

[1] Baby boomers approach 65 glumly. www.pewsocialtrends.org/2010/1 2/20/baby-boomers-approach-65-glumly/.

[2] Vincent, G. K. and Velkoff, V. A. (2010). *The Next Four Decades: The Older Population in the United States: 2010 to 2050*. Number 1138. US Department of Commerce, Economics and Statistics Administration, US Census Bureau.

[3] Carpenter, M. G., Adkin, A. L., Brawley, L. R. and Frank, J. S. (2006). Postural, physiological and psychological reactions to challenging balance: does age make a difference? *Age Ageing*, 35(3):298–303.

[4] DeVito, C. A., Lambert, D. A., Sattin, R. W., Bacchelli, S., Ros, A. and Rodriguez, J. G. (1988). Fall injuries among the elderly. community-based surveillance. *Journal of the American Geriatrics Society*, 36(3):1029–1035.

[5] Young, W. R. and Williams, A. M. (2015). How fear of falling can increase fall-risk in older adults: applying psychological theory to practical observations. *Gait Posture*, 41(1):7–12.

[6] Deslandes, A. C., Moraes, H., Alves, H., Pompeu, F. A., Silveira, H., Mouta, R., Arcoverde, C., Ribeiro, P. Cagy, M., Piedade, R. A., Laks, J. and Coutinho, E. S. (2010). Effect of aerobic training on eeg alpha asymmetry and depressive symptoms in the elderly: a 1-year follow-up study. *Brazilian Journal of Medical and Biological Research*, 43(6):585–592.

[7] Slobounov, S. M., Moss, S. A., Slobounova, E. S. and Newell, K. M. (1998). Aging and time to instability in posture. *Journal of Gerontology, Series A, Biological Sciences and Medical Sciences*, 53(1):B71–8.

[8] Brown, L. A., White, P., Doan, J. B. and de Bruin, N. (2011). Selective attentional processing to fall-relevant stimuli among older adults who fear falling. *Experimental Aging Research*, 37(3):330–45.

[9] Hadjistavropoulos, T., Delbaere, K. and Theresa Dever Fitzgerald, T. D. (2011). Reconceptualizing the role of fear of falling and balance confidence in fall risk. *Journal of Aging and Health*, 23(1):3–23.

[10] Friedman, S. M., Munoz, B., West, S. K., Rubin, G. S. and Fried, L. P. (2002). Falls and fear of falling: which comes first? a longitudinal prediction model suggests strategies for primary and secondary prevention. *Journal of American Geriatrics Society*, 50(8):1329–1335.

[11] Carpenter, M. G., Frank, J. S. and Silcher, C. P. (1999). Surface height effects on postural control: a hypothesis for a stiffness strategy for stance. *Journal of Vestibular Research*, 9(4):277–286.

[12] Whitney, S. L., Jacob, R. G., Sparto, P. J., Olshansky, E. F., Detweiler-Shostak, G., Brown, E. L. and Furman, J. M. (2005). Acrophobia and pathological height vertigo: indications for vestibular physical therapy? *Physical Therapy*, 85(5):443–458.

[13] Depla, M. F., Margreet, L., van Balkom, A. J. and de Graaf, R. (2008). Specific fears and phobias in the general population: results from the

netherlands mental health survey and incidence study (nemesis). *Social Psychiatry and Psychiatric Epidemiology*, 43(3):200–208.

[14] Grimbergen, Y. A. M., Munneke, M. and Bloem, B. R. (2004). Falls in Parkinson's disease. *Current Opinion in Neurology*, 17(4):405–415.

[15] Koller, W. C., Glatt, S., Vetere-Overfield, B. and Hassanein, R. (1989). Falls and parkinson's disease. *Clinical Neuropharmacology*, 12(2):98–105.

[16] Riva, G. (1998). From toys to brain: Virtual reality applications in neuroscience. *Virtual Reality*, 3(4):259–266.

[17] Merians, A. S., Tunik, E. and Adamovich, S. V. (2009). Virtual reality to maximize function for hand and arm rehabilitation: exploration of neural mechanisms. *Studies in Health Technology and Informatics*, 145:109.

[18] Adamovich, S. V., August, K., Merians, A. and Tunik, E. (2009). A virtual realitybased system integrated with fmri to study neural mechanisms of action observation-execution: a proof of concept study. *Restorative Neurology and Neuroscience*, 27(3):209–223.

[19] Coelho, C. M., Waters, A. M., Hine, T. J. and Wallis, G. (2009). The use of virtual reality in acrophobia research and treatment. *Journal of Anxiety Disorders*, 23(5):563–574.

[20] Eftekharsadat, B., Babaei-Ghazani, A., Mohammadzadeh, M., Talebi, M., Eslamian, F. and Azari E. (2015). Effect of virtual reality-based balance training in multiple sclerosis. *Neurological Research*, 37(6):539–544.

[21] Teo, W. P., Muthalib, M., Yamin, S., Hendy, A. M., Bramstedt, K., Kotsopoulos, E., Perrey, S. and Ayaz, H. (2016). Does a combination of virtual reality, neuromodulation and neuroimaging provide a comprehensive platform for neurorehabilitation? – a narrative review of the literature. *Frontiers in Human Neuroscience*, 10:284.

[22] Widdowson, C., Ganhotra, J., Faizal, M., Wilko, M., Parikh, S., Adhami, Z. and Hernandez, M. E. (2016). Virtual reality applications in assessing the effect of anxiety on sensorimotor integration in human postural control. In *Engineering in Medicine and Biology Society (EMBC), 2016 IEEE 38th Annual International Conference of the*, pages 33–36. IEEE.

[23] Moro, S. B., Bisconti, S., Muthalib, M., Spezialetti, M., Cutini, S., Ferrari, M., Placidi, G. and Quaresima V. (2014). A semi-immersive virtual reality incremental swing balance task activates prefrontal

cortex: a functional near-infrared spectroscopy study. *Neuroimage*, 85 Pt 1:451–460.

[24] Kaur, R., Lin, X., Layton, A., Hernandez, M. and Sowers, R. (2018). Virtual reality, visual cliffs, and movement disorders. In *2018 40th Annual International Conference of the IEEE Engineering in Medicine and Biology Society (EMBC)*, pages 81–84. IEEE.

[25] Adkin, A. L., Campbell, A. D., Chua, R. and Carpenter, M. G. (2008). The influence of postural threat on the cortical response to unpredictable and predictable postural perturbations. *Neuroscience Letters*, 435(2):120–125.

[26] Sibley, K. M., Mochizuki, G., Frank, J. S. and McIlroy, W. E. (2010). The relationship between physiological arousal and cortical and autonomic responses to postural instability. *Experimental Brain Research*, 203(3):533–540.

[27] Eftekharsadat, B., Babaei-Ghazani, A., Mohammadzadeh, M., Talebi, M., Eslamian, F. and Azari, E. (2015). Effect of virtual reality-based balance training in multiple sclerosis. *Neurol Res.*, 37(6):539–544.

[28] Peterson, S. M., Furuichi, E. and Ferris, D. P. (2018). Effects of virtual reality high heights exposure during beam-walking on physiological stress and cognitive loading. *PloS One*, 13(7):e0200306.

[29] Calabró, R.S., Naro, A., Russo, M., Leo, A., De Luca, R., Balletta, T., Buda, A., La Rosa, G., Bramanti, A. and Bramanti, P. (2017). The role of virtual reality in improving motor performance as revealed by eeg: a randomized clinical trial. *Journal of Neuroengineering and Rehabilitation*, 14(1):53.

[30] Luu, T. P., Nakagome, S., He, Y. and Contreras-Vidal, J. L. (2017). Real-time eeg-based brain-computer interface to a virtual avatar enhances cortical involvement in human treadmill walking. *Scientific Reports*, 7(1):8895.

[31] King, C. E., Wang, P. T., Chui, L. A., Do, A. H. and Nenadic, Z. (2013). Operation of a brain-computer interface walking simulator for individuals with spinal cord injury. *Journal of Neuroengineering and Rehabilitation*, 10(1):77.

[32] Wang, P. T., King, C. E., Chui, L.A., Do, A. H. and Nenadic, Z. (2012). Self-paced brain-computer interface control of ambulation in a virtual reality environment. *Journal of Neural Engineering*, 9(5):056016.

[33] Kaur, R., Sun, R., Ziegelman, L., Sowers, R. and Hernandez, M. E. (2019). "Using Virtual Reality to Examine the Neural and

Physiological Responses to Height and Perturbations in Quiet Standing," 2019 41st Annual International Conference of the IEEE Engineering in Medicine and Biology Society (EMBC), Berlin, Germany, 5233–5236.

[34] Htc vive. https://www.vive.com/us/.

[35] Brainvision eeg system. https://www.brainproducts.com/products_by_type.php?tid=1.

[36] Heart rate monitor. https://www.zephyranywhere.com/media/download/hxm1-ug-p-hxm-bt-user-guide-20130430-v01.pdf.

[37] Unity. https://unity.com/.

[38] Vizard. https://www.worldviz.com/ vizard-virtual-reality-software.

[39] Blender. https://www.blender.org/.

[40] Nvidia geforce gtx 1070. https://www.nvidia.com/en-in/geforce/products/10series/geforce-gtx-1070/.

[41] C-mill treadmill. https://www.motekforcelink.com/ product/c-mill/.

[42] Neurocom balance mnaster. https:// newborncare.natus.com/products-services/ neurocom-balance-master-systems.

[43] Python. https://www.python.org/.

[44] Mne. https://mne-tools.github.io/stable/index. html.

[45] scikit-learn. https://scikit-learn.org/stable/.

[46] pandas. https://pandas.pydata.org/.

[47] Shao, L. and Zhou, H. (1996). Curve fitting with bezier cubics. *Graphical Models and Image Processing*, 58(3):223–232.

[48] Won, H. J., Hwa, C. C. and Song, L. K. (2014). On the mathematic modeling of non-parametric curves based on cubic Bézier curves. *arXiv preprint arXiv:1411.6365*.

[49] Eeg head cap. https://www. brainproducts.com/files/public/downloads/actiCAP-64-channel-Standard-2_1201.pdf.

[50] Lai, E. *Practical Digital Signal Processing*. Elsevier, 2003.

[51] Jung, T. P., Makeig, S., Humphries, C., Lee, T. W., McKeown, M. J., Iragui, V. and Sejnowski, T. J. (200). Removing electroencephalographic artifacts by blind source separation. *Psychophysiology*, 37(2):163–78.

[52] Jung, T. P., Makeig, S., Humphries, C., Lee, T. W., McKeown, M. J., Iragui, V. and Sejnowski, T. J. (2000). Removing electroencephalographic artifacts by blind source separation. *Psychophysiology*, 37(2):163–178.

[53] Jung, T.-P., Makeig, S., Westerfield, M., Townsend, J., Courchesne, E. and Sejnowski, T. J. (2000). Removal of eye activity artifacts from

visual event-related potentials in normal and clinical subjects. *Clinical Neurophysiology*, 111(10):1745–1758.

[54] Makeig, S., Bell, A. J. Jung, T.-P. and Sejnowski, T. J. (1996). Independent component analysis of electroencephalographic data. In *Advances in Neural Information Processing Systems*, pages 145–151.

[55] Gwin, J. T. Gramann, K. Makeig, S. and Ferris, D. P. (2010). Removal of movement artifact from high-density eeg recorded during walking and running. *Journal of Neurophysiology*, 103(6):3526–3534.

[56] Landers, D. M. and Petruzzello, S. J. (1994). State anxiety reduction and exercise: does hemispheric activation reflect such changes? *Med Sci Sports Exerc.*, 26(8):1028–35.

[57] Davidson, R. J. (1998). Eeg measures of cerebral asymmetry: conceptual and methodological issues. *International Journal of Neuroscience*, 39(1–2):71–89.

[58] Widmann, A. Schroger, E. and Maess, B. (2015). Digital filter design for electrophysiological data – a practical approach. *Journal of Neuroscience Methods*, 250:34–46.

[59] Jiang, X., Bian, G.-B. and Tian, Z. (2019). Removal of artifacts from eeg signals: A review. *Sensors*, 19(5):987.

[60] Kanoga, S. and Mitsukura, Y. (2017). Review of artifact rejection methods for electroencephalographic systems. *Electroencephalography*, page 69.

[61] Kaur, J. and Kaur, A. (2015). A review on analysis of eeg signals. In *Computer Engineering and Applications (ICACEA), 2015 International Conference on Advances in*, pages 957–960. IEEE, 2015.

[62] Independent component properties image. https://labeling. ucsd.edu/ tutorial/format.

[63] Eeg independent component labeling. https://labeling. ucsd.edu/ tutorial/labels.

[64] Pion-Tonachini, L., Makeig, S. and Kreutz-Delgado, K. (2017). Crowd labeling latent dirichlet allocation. *Knowledge and Information Systems*, 53(3):749–765.

[65] Zhou, W. and Gotman, J. (2009). Automatic removal of eye movement artifacts from the eeg using ica and the dipole model. *Progress in Natural Science*, 19(9):1165–1170.

[66] Malik, M., Kim, S., Jeong, M. and Kamran, M. (2016). Hybrid eegeye tracker: Automatic identification and removal of eye movement and blink artifacts from electroencephalographic signal. *Sensors*, 16(2):241.

[67] Kong, W., Zhou, Z., Hu, S., Zhang, J., Babiloni, F. and Dai, G. (2013). Automatic and direct identification of blink components from scalp eeg. *Sensors*, 13(8):10783–10801.

[68] Radüntz, T., Scouten, J. Hochmuth, O. and Meffert, B. (2017). Automated eeg artifact elimination by applying machine learning algorithms to ica-based features. *Journal of Neural Engineering*, 14(4):046004.

[69] Tamburro, G., Fiedler, P., Stone, D., Haueisen, J. and Comani, S. (2018). A new ica-based fingerprint method for the automatic removal of physiological artifacts from eeg recordings. *PeerJ*, 6:e4380.

[70] Lin, C. T., Wu, R. C., Liang, S. F., Chao, W. H., Chen, J. L. and Jung, T. P. (2005). Eeg-based drowsiness estimation for safety driving using independent component analysis. *IEEE Transactions on Circuits and Systems I: Regular Papers*, 52(12):2726–2738.

[71] Fischer, N. L., Peres, R. and Fiorani, M. (2018). Frontal alpha asymmetry and theta oscillations associated with information sharing intention. *Frontiers in Behavioral Neuroscience*, 12, 2018.

[72] Palomba, D., Mennella, R. and Patron, E. (2017). Frontal alpha asymmetry neurofeedback for the reduction of negative affect and anxiety. *Behav Res Ther*, 92:32–40.

[73] Giesbrecht, T., Quaedflieg, C. W., Smulders, F. T., Meijer, E. H., Merckelbach, H. L. Meyer, T. and Smeets, T. (2015). The role of frontal eeg asymmetry in post-traumatic stress disorder. *Biol Psychol*, 108:62–77.

[74] Chalmers, J. A., Quintana, D. S., Abbott, M. J., Kemp, A. H. et al. (2014). Anxiety disorders are associated with reduced heart rate variability: a meta-analysis. *Frontiers in Psychiatry*, 5:80.

[75] Cohen, D. C. (1977). Comparison of self-report and overtbehavioral procedures for assessing acrophobia. *Behavior Therapy*, 8(1):17–23.

[76] Maki, B. E., Holliday, P. J. and Topper, A. K. (1991). Fear of falling and postural performance in the elderly. *Journal of Gerontology*, 46(4):M123–M131.

[77] Girden. E. R. (1992). *ANOVA: Repeated measures*. Number 84. Sage.

[78] Shapiro, S. S. and Wilk, M. B. (1965). An analysis of variance test for normality (complete samples). *Biometrika*, 52(3/4):591–611.

[79] Schultz, B. B. (1985). Levene's test for relative variation. *Systematic Zoology*, 34(4):449–456.

[80] Reason, J. T. (1978). Motion sickness adaptation: a neural mismatch model. *Journal of the Royal Society of Medicine*, 71(11):819–829.

[81] Brun, C. Gagné, M., McCabe, C. S. and Mercier, C. (2018). Motor and sensory disturbances induced by sensorimotor conflicts during passive and active movements in healthy participants. *PloS One*, 13(8):e0203206.

[82] Stevens, J. A., Corso, P. S., Finkelstein, E. A. and Miller, T. R. (2006). The costs of fatal and non-fatal falls among older adults. *Injury Prevention*, 12(5):290–295.

[83] Hartholt, K. A., Polinder, S., Van Der Cammen, T. J. M., Panneman, M. J. M., Van Der Velde, N., Van Lieshout, E. M. M., Patka, P. and Van Beeck, E. F. (2012). Costs of falls in an ageing population: A nationwide study from the Netherlands (2007–2009). *Injury*, 2012.

[84] Roudsari, B. S., Ebel, B. E., Corso, P. S., Molinari, N. A. M. and Koepsell, T. D. (2005). The acute medical care costs of fall-related injuries among the US older adults. *Injury*, 36(11):1316–1322.

[85] Court-Brown, C. M. and Caesar, B. (2006). Epidemiology of adult fractures: a review. *Injury*, 37(8):691–697.

[86] Harvey, L. A. and Close, J. C. T. (2012). Traumatic brain injury in older adults: Characteristics, causes and consequences. *Injury*.

[87] Arfken, C. L., Lach, H. W., Birge, S. J. and Miller, J. P. (1994). The prevalence and correlates of fear of falling in elderly persons living in the community. *American Journal of Public Health*.

[88] Scheffer, A. C., Schuurmans, M. J., Van dijk, N., Van der hooft, T. and De rooij, S. E. (2008). Fear of falling: Measurement strategy, prevalence, risk factors and consequences among older persons.

[89] Keshner, E. A., Streepey, J., Dhaher, Y. and Hain, T. (2007). Pairing virtual reality with dynamic posturography serves to differentiate between patients experiencing visual vertigo. *J Neuroeng Rehabil*, 4:24.

[90] Kalron, A., Fonkatz, I., Frid, L., Baransi, H. and Achiron, A. (2016). The effect of balance training on postural control in people with multiple sclerosis using the CAREN virtual reality system: A pilot randomized controlled trial. *Journal of NeuroEngineering and Rehabilitation*.

[91] Peruzzi, A., Cereatti, A., Croce, U. D. and Mirelman, A. (2016). Effects of a virtual reality and treadmill training on gait of subjects with multiple sclerosis: A pilot study. *Multiple Sclerosis and Related Disorders*.

[92] Giotakos, O., Tsirgogianni, K. and Tarnanas, I. (2007). A virtual reality exposure therapy (VRET) scenario for the reduction of fear of falling

and balance rehabilitation training of elder adults with hip fracture history. In *2007 Virtual Rehabilitation, IWVR.*

[93] Neri, S. G. R., Cardoso, J. R., Cruz, L., Lima, R. M., De Oliveira, R. J., Iversen, M. D. and Carregaro, R. L. (2017). Do virtual reality games improve mobility skills and balance measurements in community-dwelling older adults? Systematic review and meta-analysis.

[94] Duque, G., Boersma, D., Loza-Diaz, G., Hassan, S., Suarez, H., Geisinger, D., Suriyaarachchi, P., Sharma A. and Demontiero, O. (2013). Effects of balance training using a virtual-reality system in older fallers. *Clinical Interventions in Aging.*

[95] Krijn, M., Emmelkamp, P. M. G., Olafsson, R. P. and Biemond, R. (2004). Virtual reality exposure therapy of anxiety disorders: A review.

[96] Levy, F., Leboucher, P., Rautureau, G., Komano, O., Millet, B. and Jouvent, R. (2016). Fear of falling: Efficacy of virtual reality associated with serious games in elderly people. *Neuropsychiatric Disease and Treatment.*

[97] Rothbaum, B. O., Hodges, L. F., Kooper, R., Opdyke, D., Williford, J. S. and North, M. (1995). Virtual reality graded exposure in the treatment of acrophobia: A case report. *Behavior Therapy.*

[98] Rothbaum, B. O., Hodges, L., Watson, B. A. Kessler, C. D. and Opdyke, D. (1996). Virtual reality exposure therapy in the treatment of fear of flying: A case report. *Behaviour Research and Therapy.*

[99] Garcia-Palacios, A., Hoffman, H., Carlin, A., Furness, T. A. and Botella, C. (2002). Virtual reality in the treatment of spider phobia: a controlled study. *Behaviour Research and Therapy.*

[100] Rothbaum, B. O. and Hodges, L. F. (1999). The use of virtual reality exposure in the treatment of anxiety disorders.

[101] Rothbaum, B. O., Hodges, L., Smith, S., Lee, J. H. and Price, L. (2000). A controlled study of virtual reality exposure therapy for the fear of flying. *Journal of Consulting and Clinical Psychology.*

[102] Carlin, A. S., Hoffman, H. G. and Weghorst, S. (1997). Virtual reality and tactile augmentation in the treatment of spider phobia: A case report. *Behaviour Research and Therapy.*

[103] Coelho, C. M., Waters, A. M., Hine, T. J. and Wallis, G. (2009). The use of virtual reality in acrophobia research and treatment. *Journal of Anxiety Disorders*, 23(5):563–74.

[104] Klein, R. A. (2000). Virtual reality exposure therapy in the treatment of fear of flying. *Journal of Contemporary Psychotherapy.*

[105] Rus-Calafell, M. Gutiérrez-Maldonado, J., Botella, C. and Banos, R. M. (2013). Virtual reality exposure and imaginal exposure in the treatment of fear of flying: A pilot study. *Behavior Modification.*

[106] Anderson, P. L., Price, M., Edwards, S. M., Mayowa A. Obasaju, Schmertz, S. K., Zimand, E. and Calamaras, M. R. (2013). Virtual reality exposure therapy for social anxiety disorder: A randomized controlled trial. *Journal of Consulting and Clinical Psychology.*

[107] Pull, C. B. (2005). Current status of virtual reality exposure therapy in anxiety disorders: editorial review. *Current Opinion in Psychiatry.*

[108] Maples-Keller, J. L., Yasinski, C., Manjin, N. and Barbara Olasov Rothbaum, B. O. (2017). Virtual Reality-Enhanced Extinction of Phobias and Post-Traumatic Stress.

[109] Rothbaum, B. O., Anderson, P., Zimand, E., Hodges, L., Lang, D. and Wilson, J. (2006). Virtual reality exposure therapy and standard (in vivo) exposure therapy in the treatment of fear of flying. *Behavior Therapy.*

[110] Klinger, E., Bouchard, S., Legeron, P., Roy, S., Lauer, F., Chemin, I. and Nugues, P. (2005). Virtual reality therapy versus cognitive behavior therapy for social phobia: A Preliminary Controlled Study. *CyberPsychology & Behavior.*

[111] Muhlberger, A., Herrmann, M. J., Wiedemann, G., Ellgring, H. and Pauli, P. (2001). Repeated exposure of flight phobics to flights in virtual reality. *Behaviour Research and Therapy.*

[112] Gregg, L. and Tarrier, N. (2007). Virtual reality in mental health. *Social Psychiatry and Psychiatric Epidemiology.*

[113] Price, M., Mehta, N., Tone, E. B. and Anderson, P. L. (2011). Does engagement with exposure yield better outcomes? Components of presence as a predictor of treatment response for virtual reality exposure therapy for social phobia. *Journal of Anxiety Disorders.*

[114] Maltby, N. Kirsch, I. Mayers, M. and Allen, G. J. (2002). Virtual reality exposure therapy for the treatment of fear of flying: A controlled investigation. *Journal of Consulting and Clinical Psychology.*

[115] Hilfert, T. and Konig, M. (2016). Low-cost virtual reality environnment for engineering and construction. *Visualization in Engineering*, 4(1):2.

[116] Lau, T. M., Gwin, J. T. and Ferris, D. P. (2012). How many electrodes are really needed for eeg-based mobile brain imaging? *Journal of Behavioral and Brain Science*, 2(03):387.

[117] Stehlin, S. A. F., Nguyen, X. P. and Niemz, M. H. (2018). Eeg with a reduced number of electrodes: Where to detect and how

to improve visually, auditory and somatosensory evoked potentials. *Biocybernetics and Biomedical Engineering*, 38(3):700–707.

[118] Javanmard, A., Pedram Pad, P., Babaie-Zadeh, M. and Jutten, C. (2008). Estimating the mixing matrix in underdetermined sparse component analysis (sca) using consecutive independent component analysis (ica). In *Signal Processing Conference, 2008 16th European*, pages 1–5. IEEE.

[119] Kotsiantis, S. B., Zaharakis, I. and Pintelas, P. (2007). Supervised machine learning: A review of classification techniques. *Emerging Artificial Intelligence Applications in Computer Engineering*, 160:3–24.

[120] Lu, J., Behbood, V., Hao, P., Zuo, H., Xue, S. and Zhang, G. (2015). Transfer learning using computational intelligence: a survey. *Knowledge-Based Systems*, 80:14–23.

[121] Krizhevsky, A., Sutskever, I. and Hinton, G. E. (2012). Imagenet classification with deep convolutional neural networks. In *Advances in Neural Information Processing Systems*, pages 1097–1105.

[122] Takens, F. (1993). Detecting nonlinearities in stationary time series. *Int. J. Bifurcation and Chaos*, 3:241.

[123] Kwakernaak, H. and Sivan, R. (1972). *Linear Optimal Control Systems*, volume 1. Wiley-interscience New York.

[124] Rui, L., Nejati, H., Safavi, S. H. and Cheung, N.-M. (2017). Simultaneous low-rank component and graph estimation for high-dimensional graph signals: Application to brain imaging. In *Acoustics, Speech and Signal Processing (ICASSP), 2017 IEEE International Conference on*, pages 4134–4138. IEEE.

[125] Tibshirani, R., Wainwright, M. and Hastie, T. (2015). *Statistical Learning with Sparsity: the Lasso and Generalizations*. Chapman and Hall/CRC.

[126] Chung, M. K., Villalta-Gil, V., Lee, H., Rathouz, P. L. Lahey, B. B. and Zald, D. H. (2017). Exact topological inference for paired brain networks via persistent homology. In *International Conference on Information Processing in Medical Imaging*, pages 299–310. Springer.

[127] Bhattacharya, S., Ghrist, R. and Kumar, V. (2015). Persistent homology for path planning in uncertain environments. *IEEE Transactions on Robotics*, 31(3):578–590.

[128] Wu, Y., Shindnes, G., Karve, V., Yager, D., Work, D. B., Chakraborty, A. and Sowers, R. B. (2017). Congestion barcodes: Exploring the topology of urban congestion using persistent homology. In *2017 IEEE*

20th International Conference on Intelligent Transportation Systems (ITSC), pages 1–6.

[129] Liang, Z., Zhang, G., Huang, J. X. and Hu, Q. V. (2014). Deep learning for healthcare decision making with emrs. In *Bioinformatics and Biomedicine (BIBM), 2014 IEEE International Conference on*, pages 556–559. IEEE, 2014.

[130] Satone, V., Kaur, R., Faghri, F., Nalls, M. A., Singleton, A. B. and Campbell, R. H. (2018). Learning the progression and clinical subtypes of alzheimer's disease from longitudinal clinical data. *arXiv preprint arXiv:1812.00546*.

[131] Norizam, S., Taib, M. N., Lias, S., Murat, Z. H., Aris, S. A. M. and Hamid, N. H. A. (2011). Eegbased stress features using spectral centroids technique and k-nearest neighbor classifier. In *2011 UKSim 13th International Conference on Modelling and Simulation*, pages 69–74. IEEE.

[132] Alimardani, M., Keshmiri, S., Sumioka, H. and Hiraki, K. (2018). Classification of eeg signals for a hypnotrack bci system. In *2018 IEEE/RSJ International Conference on Intelligent Robots and Systems (IROS)*, pages 240–245. IEEE.

[133] Mansson, K. N. T., Frick, A., Boraxbekk, C. J., Marquand, A. F. and Williams, S. C. R. (2015). Per Carlbring, Gerhard Andersson, and Tomas Furmark. Predicting long-term outcome of internet-delivered cognitive behavior therapy for social anxiety disorder using fmri and support vector machine learning. *Translational Psychiatry*, 5(3):e530.

[134] Rasmus, A., Brudny, J., Grzelczak, M., Cysewski, P., Dziembowska, I. and Izdebski, P. (2015). Effects of heart rate variability biofeedback on eeg alpha asymmetry and anxiety symptoms in male athletes: A pilot study. *Appl Psychophysiol Biofeedback*, 41(2):141–50.

[135] Mennella, R., Patron, E. and Palomba, D. (2017). Frontal alpha asymmetry neurofeedback for the reduction of negative affect and anxiety. *Behaviour Research and Therapy*, 92:32–40.

4

Robotics in Virtual Reality

4.1 Body-in-the-Loop Control of Soft Robotic Exoskeletons During Virtual Manual Labor Tasks

Manuel E. Hernandez[1,*], Richard Sowers[2], Nicholas Thompson[2], Girish Krishnan[2] and Elizabeth T. Hsiao-Wecksler[3]

[1]Department of Kinesiology and Community Health,
University of Illinois at Urbana-Champaign, USA
[2]Department of Industrial and Enterprise Systems Engineering,
University of Illinois at Urbana-Champaign, USA
[3]Department of Mechanical Science and Engineering,
University of Illinois at Urbana-Champaign, USA
E-mail: mhernand@illinois.edu
*Corresponding Author

The combination of real-time sensing of physical and mental stress in humans during virtual manual labor tasks together with state-of-the-art active soft robotic exoskeletons holds much promise towards the development of integrated, intuitive, and responsive human-machine systems for use in stressful industrial work environments. Through the real-time monitoring of neural and physiological measures of stress arising during virtual manual labor tasks, identification of the most appropriate sensor suites for identifying mental and physical stressors is possible. Furthermore, through the integration of a person's intent, perception, and behavior in a stressful industrial setting, soft robotic exoskeletons may successfully act to alleviate physical and mental stress in a human user. However, further work remains to build an engineered system which can be of real benefit to workers in industrial settings. This review examines potential applications of integrated virtual reality and robotics in industrial settings and highlights challenges and opportunities for future human-machine systems.

4.1.1 Introduction

Excessive state anxiety or stress has been shown to have detrimental effects on physical and cognitive performance [1–4]. Even when performance is maintained, cognitive efficiency is decreased under conditions of high stress or anxiety [5–7], which increases the sense of effort during a task. While the inverted U relationship between anxiety or stress and performance has been well established [1, 2], questions remain about the underlying model that might best explain how human performance is modulated by high task accuracy demands and anxiety or stress. The processing efficiency theory [8,9] and cognitive energetic framework [10] provide a model for understanding the effect of stress on human performance but require further confirmation in actual or virtual work environments. Increased attentional bias can occur under the presence of anxiety or psychological stress [11, 12], which may be maladaptive in complex work environments and lead to an increased risk of injury. Injury itself can lead to increased stress in the human body and exacerbate stress or anxiety, which would lead to a downward spiral (Figure 4.1.1). However, the integration of both virtual reality (VR) and soft robotic exoskeleton technologies offer the potential for integrated, intuitive, and responsive human-machine systems for use in stressful industrial work environments.

Of all occupational injuries and illnesses that require days away from work in the US, approximately one third are due to work-related musculoskeletal disorders (WMSD) [13]. These WMSD resulted in a moderately high incidence rate of 29.8 cases per 10,000 full-time workers, with a median

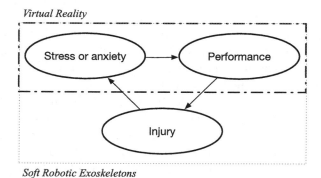

Figure 4.1.1 Concept map illustrating how stress or anxiety can impact task performance and injury, and the overlapping contributions of virtual reality or soft robotic exoskeleton technologies.

of 12 days away from work [13]. Musculoskeletal disorders are painful disorders associated with soft tissue (i.e., muscle, tendon, ligament, cartilage, nerves) [14], that can occur from either an acute bout of excessive physical effort or prolonged exposure to repetitive stress [15]. WMSD can commonly manifest as carpal tunnel syndrome, tendonitis of the shoulder, elbow, wrist or other joints, herniated discs, sciatica, or low back pain [14, 16]. In particular, occupations that have repetitive motions, high force exertion, awkward body postures, high impact force, constant vibration, and extreme temperatures pose a significant risk for WMSD [14]. Some tasks associated with WMSD are lifting and carrying heavy objects, working overhead, using handheld power-actuated fastening tools, and using high impact tools such as jack-hammers or riveters [16, 17]. Thus, devising ways to protect the worker and mitigate the effect of these risk factors is necessary to prevent WMSD and improve worker quality of life and quality of work.

Anxiety can be defined as a multi-system response to a perceived threat or danger. Anxiety brings about a variety of physical symptoms, such as cardiac acceleration, sweating, and tremors. Physiological changes arise from high stress and anxiety conditions which can be monitored in real-time [18–21]. Using frequency-specific fluctuations in heart rate dynamics, together with skin conductance or prefrontal cortical activation levels, changes in stress and anxiety can be monitored in real-time. In addition, anxiety has been linked with increased muscle co-contraction and altered body movements while performing volitional movements [22–26]. Thus, changes in movement strategies and efficiency of movement may provide mechanically intrinsic measures of anxiety in adults. However, the sensor suite which provides the most informative signal for sensing stress changes during the performance of cognitively and physically demanding tasks in the industry remains an open question.

The combination of VR and smart and active robotics holds much promise towards the development of integrated, intuitive, and responsive human-machine systems for mitigating WMSD risk factors in stressful industrial work environments. Within this review, we examine the use of robotics in alleviating WMSD risk factors in industrial settings in Section 4.1.1; the potential use of VR in evaluating the effect of stress and anxiety in industrial settings in Section 4.1.2; potential applications of integrated virtual reality and robotics in industrial settings in Section 4.1.3; and provide concluding remarks about the challenges and opportunities for future human-machine systems in Section 4.1.4.

4.1.2 Use of Robotics to Prevent Injury in Industry

A number of devices and exoskeletons have been explored to assist industrial workers to prevent injury or to augment workers to allow them to perform enhanced tasks. Equipois, Lockheed Martin, and EXHAUSS have created commercially-available passive, unpowered mechanical arms for transferring the load of heavy tools, up to 16.3 kg, away from the arms through external support structures that are attached, or not attached to the worker [27–29]. Otherlab (which recently spun off as Roam Robotics) has proposed an inflatable pneumatic exoskeleton for knee support during heavy lifting [30, 31]. The devices and exoskeletons are large, heavy, rigid and, if powered, require complex control for precision movement. All, including the pneumatic knee support, are quite bulky. Furthermore, it is unclear if any device helps to mitigate the effects of long-term exposure to tool vibration. Thus, there is a clear need for wearable smart personal protective equipment designs that sense when to protect workers from internal injury due to exposure to overexertion, repetitive motions, and vibration, while packaged in a lightweight and unobtrusive design.

Soft robots have become increasingly popular because of their unique advantages such as adaptability and safe interactions with the environment with minimal controls [32–34]. They differ from conventional robots as they lack rigid and stiff members, actuators and sensors. Soft robots are usually bioinspired and are comprised of stretchable skins, tendons, pressurized fluids, soft muscles, and fibers. Popular actuation schemes of soft robots include hydraulics and pneumatics [35–37], shape memory alloy actuation [38], and driving cables using motors [39]. The soft and flexible nature of these structural and actuation members enable direct translation of soft robotic technologies to wearable devices [40]. Thus, the combination of soft robotic exoskeleton technology with real-time monitoring of neural and physiological measures of stress arising during virtual manual labor tasks may successfully alleviate stress and anxiety in high demanding work environments.

4.1.3 Use of Virtual Reality in Evaluating the Effect of Stress and Anxiety in Industrial Settings

Through the use of VR, individuals can be safely exposed to conditions of increased threat to examine the role of anxiety [41] or stress [42] and be presented with precisely timed complex visual stimuli to enhance the sensitivity and reliability of experimental results. VR is capable of inducing behavioral, psychological, and electrophysiological effects comparable to real-world manipulations [26, 42] while avoiding the associated risks to health and safety. VR provides the ability to recreate training opportunities that would otherwise be too hazardous [43, 44] or too infrequent [45]. VR training of demanding tasks in industry have been shown to provide comparable results to real physical training [46] and improvements over traditional computer simulations [47] in certain conditions. In industry settings, VR may be used to monitor the behavioral patterns of workers, and identify behaviors that increase the risk to WMSD [48]. In addition, VR offers an opportunity to merge the best practices in occupational injury prevention with human-machine systems during both development or training [49, 50]. In particular, recreating physically demanding manipulation tasks in spatially complex environments may provide improvements in future deployments of human-machine systems (See Figure 4.1.2).

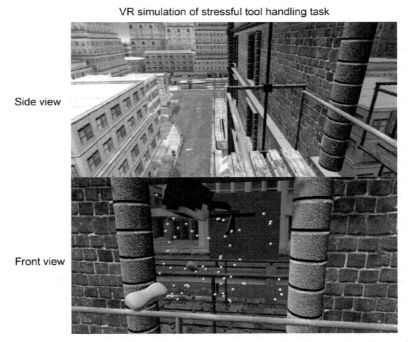

Figure 4.1.2 Side and front view of VR simulation of the stressful tool handling tasks in complex environment.

4.1.4 Applications of Integrated Virtual Reality and Robotics in Industry

4.1.4.1 Body-in-the-loop control of human-machine systems

The real-time sensing of physical and mental stress in humans during virtual manual labor tasks using mobile neuroimaging modalities together with state-of-the-art active soft robotic exoskeletons holds much promise towards the development of integrated, intuitive, and responsive human-machine systems for use in stressful industrial work environments. Through the use of electroencephalography (EEG) or functional near-infrared spectroscopy (fNIRS), changes in neural activity or oxygenation levels in the cortex can be monitored during the performance of physical tasks [51–60]. Using a combination of virtual reality and actuated feedback, instantaneously adaptable work-related scenarios may be displayed to further our understanding of the worker's response to these stimuli. Given the expected changes in brain activation patterns due to anxiety and stress, the use of EEG or fNIRS provides an objective quantification of attentional demands, which can complement other physiological measures such as skin conductance, muscle force, and heart rate variability. Together with measurements of human motion recorded via standard motion capture methods and portable inertial measurement units, these sensors can be used to provide information about real-time changes in mental and physical stress in manual labor tasks in industry [50].

4.1.4.2 Soft robotic exoskeleton actuation

Dr. Krishnan has developed an arsenal of soft fluidic actuators known as Fiber Reinforced Elastomeric Enclosures (FREEs). FREEs are composite equivalents of the popular McKibben muscles [36, 61, 62] (Figure 4.1.3(a)), which are inspired by the structure of muscular hydrostats, trees and worms [63, 64]. A FREE consists of a core hollow elastomeric cylinder reinforced by at least two families of inextensible helical fibers (Figure 4.1.3(b)). The FREE spatially deforms when the elastomeric core is pressurized with fluids. By varying the orientation of these fibers, Dr. Krishnan has reported several modes such as bending (Figure 4.1.3(c)), axial rotation (Figure 4.1.3(d)), spiral motion (Figure 4.1.3(e)) and other continuum freeform shapes (Figure 4.1.3(f)) [61, 62, 65]. Most of these deformation shapes are a result of combining FREEs with asymmetric fiber angle $\alpha 6 = -\beta$ [65], while typical engineering usage and available literature has been limited to the McKibben fiber configurations of $\alpha = -\beta$. Ongoing work [66] has resulted in a generalized

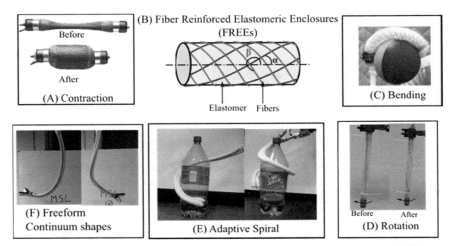

Figure 4.1.3 Fiber Reinforced Elastomeric Enclosures (FREEs) and their spatial deformation patterns.

virtual work-based formulation to predict exact deformation and actuation forces/torques as a function of applied fluid pressure and hyperelastic material properties. For a contracting McKibben actuator, the analysis reveals a tradeoff between actuation forces and energy density the former increasing nonlinearly, while the latter decreasing with FREE diameter. This tradeoff is resonated by existing state-of-the-art wearable suits [39] which though effective, are too large to be inconspicuous under clothing.

Novel methods to overcome the loss in performance of FREE-based actuators when miniaturized are needed. These would benefit from novel architectures of FREEs, which simultaneously increase actuation stroke and stiffness in the joint. These would translate to higher load-bearing ability and larger actuation displacements, beyond those permitted by using miniaturized FREEs, yet retain the benefits associated with wearability such as compactness and discreetness.

4.1.4.3 FREE architectures for stiffness modulation

Modulating stiffness increases load-bearing ability by providing support against an external load. We present architecture of FREEs for increasing the stiffness at joints based on antagonistic muscle activation in humans. For convenience, we shall demonstrate this on a one-degree of freedom elbow joint. The architecture aims to conform itself to the

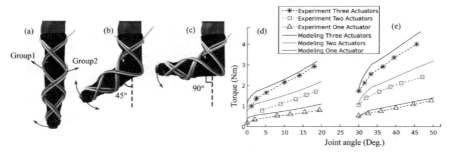

Figure 4.1.4   (a–c) Preliminary prototype and demonstration of stiffness modulation for a 1-DOF-elbow joint, (d) benchtop stiffness tests on the joint at $0°$ orientation. (e) benchtop testing of the joint at $30°$ orientation.

contours of the human arm around the elbow joint at any joint orientation. Once confirmed, sufficient external forces are required to deform the joint.

The architecture consists of two groups of helically wrapped contracting FREE actuators (Figure 4.1.4(a)). Each group is a double helix, consisting of equal and opposite wraps. Group 1 (green shown in Figure 4.1.4(a)) wraps around the anterior end of the joint, while Group 2 (red in Figure 4.1.4(a)) wraps on the posterior end. When not actuated, the FREEs loosely fit on the arm and do not affect elbow rotation. Now, by rotating the lower arm by an arbitrary angle (for example, $45°$ shown in Figure 4.1.4(b) or $90°$ in Figure 4.1.4(c)) and then pressurizing the Group 1 family of FREEs causes them to conform to the shape of the rotating arm, along the anterior end of the joint. Any joint extension tends to stretch the Group 1 FREEs and is thus resisted. This resistance is manifested as increased stiffness or support against the external load. However, joint flexion is unrestricted as it leads to a decrease in Group 1 FREE length, which is resolved by buckling. Now, pressurizing Group 2 FREEs (red) conforms them to the posterior end, thus offering stiffness against flexion. Simultaneously pressurizing both groups will lead to increased stiffness against extension and flexion.

In ongoing work, we proposed a simple analysis method to capture the increase in stiffness [39] of the sleeve using the force obtained from a single FREE and modeling the friction between the FREE and the arm. A 3D printed model of the arm was fabricated for experimental evaluation. The stiffness of the arm was tested by applying an end load leading to a moment at the joint at two initial orientations of $0°$ (or straight)

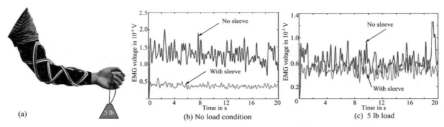

Figure 4.1.5 Preliminary testing of muscle activation levels (EMG) in biceps, under (b) no load, and (c) 5 lb load.

arm and a flexion of $30°$. Torque vs. deflection were measured and plotted in Figure 4.1.4(d,e) by activating different numbers of muscles per group (for example, there are two muscles/groups in Figure 4.1.4(a)). As expected, resistance increased proportionally with number of muscles actuated.

With benchtop evidence of successfully providing static support, a preliminary wearable prototype was fabricated and customized for an individual's arm as shown in Figure 4.1.5(a). EMG electrodes were inserted under the sleeve for measuring muscle activation levels of the biceps under two loading conditions (i) no load, (ii) hand holding 5 lb. For both cases, the arm was flexed by about $85°$ about the elbow. The EMG data collected for the no-load condition (Figure 4.1.5(b)) shows a 70% reduction in the voltage when wearing the sleeve than without. This implies that the sleeve is clearly effective in offsetting the muscle power required to overcome the arms inertia. However, wearing the sleeve showed an improvement of 20% with a weight of 5 lbs carried by the hand. Thus, we propose scaling of the present design coupled with actuation to increase the load-bearing ability of the exoskeleton, and help reduce physiological effort.

4.1.4.4 Nested FREE architectures for actuating joints

Apart from stiffening, providing even modest actuation forces on the joint may be beneficial in alleviating physical stress in manual labor tasks. Joint actuation can be achieved by placing linear actuators such as Mckibben muscles or FREEs along the forearm and the upper arm as shown in Figure 4.1.6(b). The linear actuator contracts, which creates a moment about the joint thus causing it to actuate. The angular displacement of the joint thus depends on actuator contraction or stroke, which for a FREE muscle is

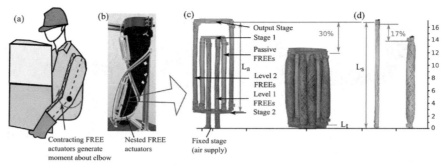

Figure 4.1.6   Nested contracting FREEs for joint actuation. (a) A single contracting actuator placed along the arm, (b) A demonstration of elbow actuation by nested FREE actuators, (c) construction and operating principle of nested architecture demonstrating significantly larger contraction ratios than (d) a single contracting FREE.

directly proportional to its length [67,68]. Our preliminary analysis reveals the need for a muscle with unusually large contraction ratio (>40% of its length) to actuate an elbow joint by 900. Furthermore, sufficient forces (our analysis predicts 40 N) must be produced to lift the arm against its own weight. This can be difficult to produce with miniaturized FREE muscles since the elastic energy stored in the bladder limits the obtained stroke, and forces [69].

4.1.5 Conclusions

In conclusion, we find that virtual reality may offer an opportunity for a better understanding of physical and mental stress in humans during manual labor tasks. VR has been used to recreate stressful physical or mental tasks in industry, which allows for novel training paradigms and real-time sensing of stress in humans during virtual manual labor tasks. Together with state-of-the-art active soft robotic exoskeletons these technologies hold much promise towards development of integrated, intuitive, and responsive human-machine systems for use in stressful industrial work environments. However, before a person's intent, perception, and behavior in a stressful industrial setting can be integrated with future human-machine systems to create engineered system which can be of real benefit to workers in industrial settings, further advances are needed to validate multimodal sensor data and soft robotic exoskeleton actuation.

Acknowledgments

We thank members of the Monolithic Systems Lab, Human Dynamics and Control Laboratory, Mobility and Fall Prevention Lab, and Illinois Geometry Lab for their assistance with discussion and materials for this review.

References

[1] R. Martens and D. Landers, "Motor performance under stress: a test of the inverted-u hypothesis," J Pers Soc Psychol, vol. 16, no. 1, pp. 29–37, 1970.

[2] A. Carron, "Reactions to "anxiety and motor behavior"," J Mot Behav, vol. 3, no. 2, pp. 181–8, 1971.

[3] J. Noteboom, K. Barnholt, and R. Enoka, "Activation of the arousal response and impairment of performance increase with anxiety and stressor intensity," J Appl Physiol (1985), vol. 91, no. 5, pp. 2093–101, 2001.

[4] R. McHugh, E. Behar, C. Gutner, D. Geem, and M. Otto, "Cortisol, stress, and attentional bias toward threat," Anxiety Stress Coping, vol. 23, no. 5, pp. 529–45, 2010.

[5] L. Hardy, "Stress, anxiety and performance," J Sci Med Sport, vol. 2, no. 3, pp. 227–33, 1999.

[6] R. Mullen and L. Hardy, "State anxiety and motor performance: testing the conscious processing hypothesis," J Sports Sci, vol. 18, no. 10, pp. 785–99, 2000.

[7] K. Mandrick, V. Peysakhovich, F. Rmy, E. Lepron, and M. Causse, "Neural and psychophysiological correlates of human performance under stress and high mental workload," Biol Psychol, vol. 121, no. Pt A, pp. 62–73, 2016.

[8] M. Eysenck, N. Derakshan, R. Santos, and M. Calvo, "Anxiety and cognitive performance: attentional control theory," Emotion, vol. 7, no. 2, pp. 336–53, 2007.

[9] N. Derakshan, T. Ansari, M. Hansard, L. Shoker, and M. Eysenck, "Anxiety, inhibition, efficiency, and effectiveness. an investigation using antisaccade task," Exp Psychol, vol. 56, no. 1, pp. 48–55, 2009.

[10] G. Hockey, "Compensatory control in the regulation of human performance under stress and high workload; a cognitive-energetical framework," Biol Psychol, vol. 45, no. 1–3, pp. 73–93, 1997.

[11] H. Miranda, E. Viikari-Juntura, R. Martikainen, E. Takala, and H. Riihimki, "Physical exercise and musculoskeletal pain among forest industry workers," Scand J Med Sci Sports, vol. 11, no. 4, pp. 239–46, 2001.

[12] M. El Khoury-Malhame, E. Reynaud, A. Soriano, K. Michael, P. Salgado-Pineda, X. Zendjidjian, C. Gellato, F. Eric, M. Lefebvre, F. Rouby, J. Samuelian, J. Anton, O. Blin, and S. Khalfa, "Amygdala activity correlates with attentional bias in ptsd," Neuropsychologia, vol. 49, no. 7, pp. 1969–73, 2011.

[13] U.S. Bureau of Labor Statistics, "Nonfatal occupational injuries and illnesses requiring days away from work," U.S. Bureau of Labor Statistics, Tech. Rep., 2016.

[14] D. Wang, F. Dai, and X. Ning, "Risk assessment of work-related musculoskeletal disorders in construction: State-of-the- art review," J Constr Eng Manage, vol. 141, no. 6, pp. 1–15, 2015.

[15] N. Inyang, M. Al-Hussein, M. El-Rich, and S. Al-Jibouri, "Ergonomic analysis and the need for its integration for planning and assessing construction tasks," J Constr Eng Manage, vol. 138, no. 12, pp. 1370–1376, 2012.

[16] J. Albers and C. Estill, "Simple solutions: Ergonomics for construction workers," United States Department of Health and Human Services, NIOSH, Tech. Rep., 2007.

[17] M. Hagberg, "Clinical assessment of musculoskeletal disorders in workers exposed to hand-arm vibration," Int Arch Occup Environ Health, vol. 75, no. 1–2, pp. 97–105, 2001.

[18] K. Lee, K. Yoon, M. Ha, J. Park, S. Cho, and D. Kang, "Heart rate variability and urinary catecholamines from job stress in korean male manufacturing workers according to work seniority," Ind Health, vol. 48, no. 3, pp. 331–8, 2010.

[19] M. Mateo, C. Blasco-Lafarga, I. Martnez-Navarro, J. Guzmn, and M. Zabala, "Heart rate variability and pre-competitive anxiety in bmx discipline," Eur J Appl Physiol, vol. 112, no. 1, pp. 113–23, 2012.

[20] A. Williamon, L. Aufegger, D. Wasley, D. Looney, and D. Mandic, "Complexity of physiological responses decreases in high-stress musical performance," J R Soc Interface, vol. 10, no. 89, p. 20130719, 2013.

[21] M. Garca-Rubio, L. Espn, V. Hidalgo, A. Salvador, and J. Gmez-Amor, "Autonomic markers associated with generalized social phobia

symptoms: heart rate variability and salivary alpha-amylase," Stress, vol. 20, no. 1, pp. 44–51, 2017.

[22] C. Adler, D. Crews, J. Hentz, A. Smith, and J. Caviness, "Abnormal co-contraction in yips-affected but not unaffected golfers: evidence for focal dystonia," Neurology, vol. 64, no. 10, pp. 1813–4, 2005.

[23] M. Yoshie, K. Kudo, T. Murakoshi, and T. Ohtsuki, "Music performance anxiety in skilled pianists: effects of social- evaluative performance situation on subjective, autonomic, and electromyographic reactions," Exp Brain Res, vol. 199, no. 2, pp. 117–26, 2009.

[24] M. Wuehr, G. Kugler, R. Schniepp, M. Eckl, C. Pradhan, K. Jahn, D. Huppert, and T. Brandt, "Balance control and anti-gravity muscle activity during the experience of fear at heights," Physiol Rep, vol. 2, no. 2, p. e00232, 2014.

[25] T. Brandt, G. Kugler, R. Schniepp, M. Wuehr, and D. Huppert, "Acro-phobia impairs visual exploration and balance during standing and walking," Ann N Y Acad Sci, vol. 1343, pp. 37– 48, 2015.

[26] T. Cleworth, R. Chua, J. Inglis, and M. Carpenter, "Influence of virtual height exposure on postural reactions to support surface translations," Gait Posture, vol. 47, pp. 96–102, 2016.

[27] P. J. Goldman D, Kelly H, "36 coolest gadgets of 2014. cnn money." https://money.cnn.com/gallery/technology/innovationnation/2014/12/10/coolest-gadgets-2014/24.html, accessed: 2018-04-14.

[28] K. D, "Nab 2015: Exhauss exoskeleton in abelcine," http://blog.abe lcine.com/2015/04/14/nab-2015-exhauss-exoskeleton/,accessed: 2018-04-14.

[29] "zerog and x-ar," http://www.equipoisllc.com/products/, accessed: 2018-04-14.

[30] "Otherlab," http://www.otherlab.com, accessed: 2018-04-14.

[31] "Roam robotics," http://www.roamrobotics.com, accessed: 2018-04-14.

[32] F. Ilievski, A. Mazzeo, X. Shepherd, RF. Chen, and G. White-sides, "Soft robotics for chemists," Angew. Chem. Int. Ed., vol. 50, no. 8, pp. 1890–1895, 2011.

[33] C. Laschi, "Soft robotics research, challenges, and innovation potential, through showcases," in Soft Robotics: Transferring Theory to Applica-tion, A. Verl, A. Albu-Schffer, O. Brock, and A. Raatz, Eds. Springer Berlin Heidelberg, 2015, pp. 225–264.

[34] D. Rus and M. Tolley, "Design, fabrication and control of soft robots," Nature, vol. 521, no. 7553, p. 467475, 2015.

[35] K. Aschenbeck, N. Kern, R. Bachmann, and R. Quinn, "Design of a quadruped robot driven by air muscles," The First IEEE/RAS-EMBS International Conference on Biomedical Robotics and Biomechatronics, BioRob 2006, vol. 24, no. 1, p. 875880, 2006.

[36] J. Bishop-Moser, G. Krishnan, and S. Kota, "Force and hydraulic displacement amplification of fiber reinforced soft actuators," Proceedings of the ASME Design Engineering Technical Conference, p. 6 A, 2013.

[37] S. Neppalli, "Octarm-a soft robotic manipulator," Intelligent Robots and Systems, 2007. IROS 2007. IEEE/RSJ International Conference on, p. 2569, 2007.

[38] S. Seok, C. Onal, K.-J. Cho, R. Wood, R. Rus, and S. Kim, "Meshworm: A peristaltic soft robot with antagonistic nickel titanium coil actuators," IEEE/ASME Trans. Mechatronics, vol. 18, no. 5, p. 14851497, 2013.

[39] M. Wehner, B. Quinlivan, P. Aubin, E. Martinez-Villalpando, M. Baumann, L. Stirling, K. Holt, R. Wood, and C. Walsh, "A lightweight soft exosuit for gait assistance," Robotics and Automation (ICRA), 2013 IEEE International Conference on, p. 33623369, 2013.

[40] P. Polygerinos, Z. Wang, K. Galloway, R. Wood, and C. Walsh, "Soft robotic glove for combined assistance and at-home rehabilitation," Rob. Auton. Syst., vol. 73, pp. 135–143, 2015.

[41] D. Yelshyna, M. Gago, E. Bicho, V. Fernandes, N. Gago, L. Costa, H. Silva, M. Rodrigues, L. Rocha, and N. Sousa, "Compensatory postural adjustments in parkinson's disease assessed via a virtual reality environment," Behav Brain Res, vol. 296, pp. 384–92, 2016.

[42] M. Groer, R. Murphy, W. Bunnell, K. Salomon, J. Van Eepoel, B. Rankin, K. White, and C. Bykowski, "Salivary measures of stress and immunity in police officers engaged in simulated critical incident scenarios," J Occup Environ Med, vol. 52, no. 6, pp. 595–602, 2010.

[43] M. Filigenzi, T. Orr, and T. Ruff, "Virtual reality for mine safety training," Appl Occup Environ Hyg, vol. 15, no. 6, pp. 465–9, 2000.

[44] I. Szke, M. Louka, T. Bryntesen, J. Bratteli, S. Edvardsen, K. REitrheim, and K. Bodor, "Real-time 3d radiation risk assessment supporting simulation of work in nuclear environments," J Radiol Prot, vol. 34, no. 2, pp. 389–416, 2014.

[45] R. Leitch, G. Moses, and H. Magee, "Simulation and the future of military medicine" Mil Med, vol. 167, no. 4, pp. 350–4, 2002.

[46] F. Ganier, C. Hoareau, and J. Tisseau, "Evaluation of procedural learning transfer from a virtual environment to a real situation: a case study on

tank maintenance training," Ergonomics, vol. 57, no. 6, pp. 828–43, 2014.

[47] J. Vora, S. Nair, A. Gramopadhye, A. Duchowski, B. Melloy, and B. Kanki, "Using virtual reality technology for aircraft visual inspection training: presence and comparison studies," Appl Ergon, vol. 33, no. 6, pp. 559–70, 2002.

[48] L. Nedel, V. de Souza, A. Menin, L. Sebben, J. Oliveira, F. Faria, and A. Maciel, "Using immersive virtual reality to reduce work accidents in developing countries," IEEE Comput Graph Appl, vol. 36, no. 2, pp. 36–46, 2016.

[49] S. Aromaa and K. Vnnen, "Suitability of virtual prototypes to support human factors/ergonomics evaluation during the design," Appl Ergon, vol. 56, pp. 11–8, 2016.

[50] A. Zimmer, "The essential tension of human-machine systems," Orthopade, vol. 31, no. 10, pp. 981–6, 2002.

[51] I. Miyai, H. Tanabe, I. Sase, H. Eda, I. Oda, I. Konishi, Y. Tsunazawa, T. Suzuki, T. Yanagida, and K. Kubota, "Cortical mapping of gait in humans: a near-infrared spectroscopic topography study," Neuroimage, vol. 14, no. 5, pp. 1186–92, 2001.

[52] H. Atsumori, M. Kiguchi, T. Katura, T. Funane, A. Obata, H. Sato, T. Manaka, M. Iwamoto, A. Maki, H. Koizumi, and K. Kubota, "Non-invasive imaging of prefrontal activation during attention-demanding tasks performed while walking using a wearable optical topography system," J Biomed Opt, vol. 15, no. 4, p. 046002, 2010.

[53] J. Gwin, K. Gramann, S. Makeig, and D. Ferris, "Electrocortical activity is coupled to gait cycle phase during treadmill walking," Neuroimage, vol. 54, no. 2, pp. 1289–96, 2011.

[54] A. Sipp, J. Gwin, S. Makeig, and D. Ferris, "Loss of balance during balance beam walking elicits a multifocal theta band electrocortical response," J Neurophysiol, vol. 110, no. 9, pp. 2050–60, 2013.

[55] J. Bradford, J. Lukos, and D. Ferris, "Electrocortical activity distinguishes between uphill and level walking in humans," J Neurophysiol, vol. 115, no. 2, p. jn.00089.2015, 2015.

[56] M. Hernandez, R. Holtzer, G. Chaparro, K. Jean, J. Balto, B. Sandroff, M. Izzetoglu, and R. Motl, "Brain activation changes during locomotion in middle-aged to older adults with multiple sclerosis," J Neurol Sci, vol. 370, pp. 277–83, 2016.

[57] G. Chaparro, J. Balto, B. Sandroff, R. Holtzer, M. Izzetoglu, R. Motl, and M. Hernandez, "Frontal brain activation changes due to dual-tasking

under partial body weight support conditions in older adults with multiple sclerosis," J Neuroeng Rehabil, vol. 14, no. 1, p. 65, 2017.

[58] R. Kaur, X. Lin, A. Layton, M. Hernandez, and R. Sowers, "Virtual reality, visual cliffs, and movement disorders," in 2018 40th Annual International Conference of the IEEE Engineering in Medicine and Biology Society (EMBC). IEEE, 2018, pp. 81–84.

[59] M. Hernandez, E. O'Donnell, G. Chaparro, R. Holtzer, M. Izzetoglu, B. Sandroff, and R. Motl, "Brain activation changes during balance and attention demanding tasks in middle and older-aged adults with multiple sclerosis," Motor Control, vol. In Press, pp. 1–20, 2019.

[60] R. Kaur*, R. Sun*, L. Ziegelman, R. Sowers, and M. Hernandez, "Using virtual reality to examine the neural and physiological responses to height and perturbations in quiet standing," in 2019 41st Annual International Conference of the IEEE Engineering in Medicine & Biology Society (EMBC). In press.

[61] J. Bishop-Moser and S. Kota, "Towards snake-like soft robots: Design of fluidic fiber-reinforced elastomeric helical manipulators," Intelligent Robots and Systems (IROS), 2013 IEEE/RSJ International Conference on, pp. 5021–5026, 2013.

[62] J. Bishop-Moser, G. Krishnan, C. Kim, and S. Kota, "Design of soft robotic actuators using fluid-filled fiber-reinforced elastomeric enclosures in parallel combinations," IEEE International Conference on Intelligent Robots and Systems, p. 42644269, 2012.

[63] S. Vogel, Comparative biomechanics: lifes physical world. Princeton University Press, 2013.

[64] S. Wainwright, Axis and Circumference: The cylindrical shape of plants and animals. Harvard Press, 1988.

[65] G. Krishnan, C. Bishop-Moser, J. Kim, and S. Kota, "Kinematics of a generalized class of pneumatic artificial muscles," J. Mech. Robot., vol. 7, no. 4, p. 41014, 2015.

[66] G. Singh and G. Krishnan, "A constrained maximization formulation to analyze deformation of fiber reinforced elastomeric actuators," Smart Mater. Struct., vol. 26, no. 6, p. 65024, 2017.

[67] C.-P. Chou and B. Hannaford, "Measurement and modeling of mckibben pneumatic artificial muscles," IEEE Transactions on Robotics and Automation, vol. 12, pp. 90–102, 1996.

[68] X. Zhang, G. Singh, and G. Krishnan, "A soft wearable sleeve for joint stiffness modulation," Advanced Intelligent Mechatronics (AIM), 2016 IEEE International Conference on. IEEE, pp. 264–269, 2016.

[69] T. Pillsbury, C. Kothera, and N. Wereley, "Effect of bladder wall thickness on miniature pneumatic artificial muscle performance," Bioinspiration and Biomimetics, vol. 10, p. 55006, 2016.

4.2 Towards Mixed Reality System with Quadrotor: Autonomous Drone Positioning in Real and Virtual

German Espinosa and Michael Rubenstein

Electrical Engineering and Computer Science Department,
Northwestern University, USA

Three-dimensional (3D) positioning in real and virtual spaces is needed for mixed reality robotic systems. In normal conditions achieving 3D positioning can be done by using traditional global positioning system (GPS). However, in confined spaces these systems constraints such as reception dropout to localization error and noise, make them unpractical or impossible to implement. We introduce a cheap, high precision, 3D positioning system leveraging the HTC Vive^{TM} virtual reality (VR) lighthouse that allows a drone to autonomously fly in real space as well as virtual space. We built and tested the new system to determine feasibility and accuracy.

4.2.1 Introduction

Interactions between objects in real and virtual spaces are a central piece in the VR Ecosystem. Having the ability to blend items from the real world into a virtual environment allows systems to perform more complex and valuable tasks using VR tools and also to provide a more immersive and rich experience. To do so, VR systems rely on high accuracy 3D positioning, without the ability to consistently and accurately locate an object in the real world it becomes impossible to program any interaction between this object and a functionality in a virtual environment. This means, the mix of reality and virtual reality can only be achieved by having detailed information of the changes in location from every item existing in both ends.

Traditionally 3D positioning has been solved using a wide range of solutions, from global satellite systems [1] to sound waves [2], WIFI signal [3], and computer vision techniques [4]. These technologies had allowed systems with multiple devices to perform synchronized tasks and navigate complex geometries. However, when applied to reality and VR blend, there are some constraints that made them unpractical or impossible to implement. For example: GPS will not perform consistently in indoors settings and the standard error margin will prevent any fine interaction from working. Using WIFI signal and sound waves for positioning works well indoors but accuracy and external contamination made them impractical and poorly effective. In the other hand, computational complexity, environment preparation and a slow refresh rate make using a computer vision system with landmarks extremely complex to implement in small robotic devices.

We propose the use the HTC Vive Lighthouse [5] on reduced spaces to obtain consistent and accurate positioning with low processing cost and at a fast refresh rate. The system relies on infrared laser beams and infrared sensors to work, which prevents most of the contamination problems and increase accuracy. We designed and built the necessary electronics to evaluate the concept and tested the final device on a flying drone holding position in both, real and virtual spaces.

4.2.2 The Vive System

The HTC Vive Lighthouse [5] contains two sets of IR light emitting devices (Figure 4.2.1). First, an omnidirectional IR flash used to synchronize all the devices receiving positioning information and second a high precision IR laser working together with a high-speed moving mirror to perform scans in two axes.

A high-level description of the interaction between the lighthouse and the devices is as follow: The Lighthouse emits a series of omnidirectional

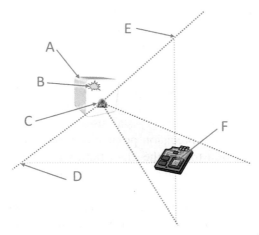

Figure 4.2.1 (A) Vive Lighthouse. (B) Omnidirectional IR Flash. (C) High speed mirror and laser beam. (D) Y axis value. (E) X axis value. (F) Vive sensor.

synchronization flashes, then it starts the scan in the x axis using the laser bean, when it finishes it emits another omnidirectional flash, and to finalize then proceeds with the y axis. This process is repeated thirty times per second.

In the other end of the HTC Vive system, the device receiving the localization information waits until a synchronization flash is detected. As the sensors cannot differentiate the different type of signals, the device measures the time between signals to determine if it is, in fact, a synchronization pulse or not. Once a sync flash is detected, the device expects a series of signals from the lighthouse in a specific order. The resulting 2D position of the sensor is estimated by computing the time difference between the flashes and the corresponding laser beam signals. Objects in the real space need to gather 2D information from multiple sensors to avoid ambiguity and effectively determine their location in a 3D space [6].

4.2.3 The Device

To simplify and encapsulate the complexity of the interaction with the HTC Vive system, we designed a lightweight and cheap device with four coplanar HTC Vive sensors, responsible of synchronizing with the lighthouse and computing the 2D position of each sensor.

The device counts with a PIC32 microcontroller that is interrupted every time a sensor receives a signal from the lighthouse. Using an 8 MHz crystal oscillator, the device times the signals received to differentiate sync flashes from positioning laser beam pulses. Once all the signals from all four sensors are received and the 2D data decoded (time to xy axis values), the information is transmitted using standard UART serial communication.

4.2.4 From 2D to 3D

After the device transmits enough 2D information, it is possible estimate the 3D pose of the object relative to the lighthouse. To solve the problem, a computer vision algorithm [7] was used with the 2D location of each sensor acting as a high precision point in a virtual image. To complete the necessary information to run the algorithm, we created a 3D model of the device specifying the location of each one of sensors in it and then computed its full pose. The implementation resulted it in a C program that reads UART to receive information from the device and process the pose. We successfully ran it in a Raspberry PI Zero W, to ensure it could be added to a flying device later.

4.2.5 Results

After the device and the program started running together, we tested the accuracy of the pose estimation using only four degrees of freedom: X, Y, Z and YAW and computed the results.

To determine yaw accuracy, 2000 measurements were performed on three different distances, 1000 with the device facing forward and 1000 with the device rotated 90 degrees. We were able to obtain the device yaw information within 2 degrees error 99% of the time (Figure 4.2.2).

In terms of 3D Position, as the device was designed in a planar structure, the accuracy and, specially, the noise of the information originated in the X and Y axis (plane used to locate the sensors) were significantly better than the information originated in the Z axis. After performing 1000 measurements on two different distances we were able to obtain Z axis positioning within a 5 cm error 98% of the time (Figure 4.2.3). However, when obtaining information on the X and Y axis, the error and noise reduced significantly. On the same measurements we obtained values within 1 cm error 98.75% of the time (Figure 4.2.4).

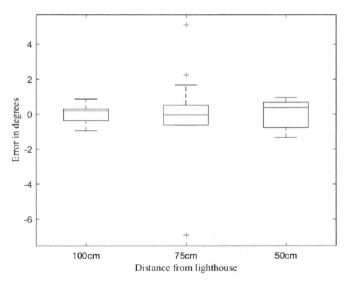

Figure 4.2.2 Yaw error boxplot, by distance. Computed from 6000 measurements, 2000 in each distance, 50% facing forward (0 degrees) and 50% facing sideways (90 degrees).

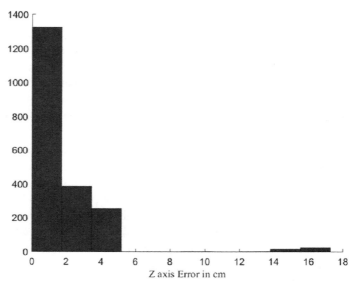

Figure 4.2.3   Z axis error histogram in cm. Computed from 2000 measurements, 50% at 100 cm and 50% at 75 cm from the lighthouse.

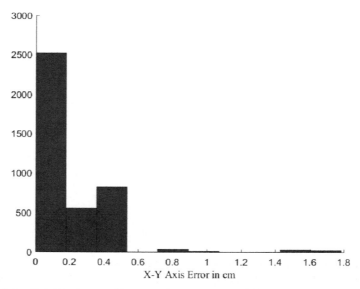

Figure 4.2.4   X & Y axis error histogram in cm. Computed from 2000 measurements, 50% at 100 cm and 50% at 75 cm from the lighthouse.

4.2.6 In Action

Once the preliminary results were satisfactory, the next step was to the test the sensor on a flying device. To do so, we mounted a Vive lighthouse on the ceiling facing down and attached the sensor to a quadcopter (Figure 4.2.5). The goal was to perform a series of autonomous flights simultaneously computing the position of the device in a virtual space.

To autonomously control the drone effectively, sensor aggregation was added using higher refresh rate data coming from the quadcopter accelerometer (Figure 4.2.6). This allowed to sense and evaluated speed information (by integrating accelerations) hundreds of times per seconds, in order of magnitude quicker than what our device can do. We used a complementary filter aggregating both sources, Vive device and IMU, to have current, accurate and reliable information.

After all positioning and velocity information was obtained, a PID [8] flight planner processes it and outputs a set of high-level instructions to affect the drone behavior in four degrees of freedom: roll, pith, yaw and thrust. After that the drone controller takes over and implements the new flight parameters.

With this implementation the quadcopter successfully maintained a stationary position in a virtual box during an autonomously controlled flight while reporting its 3D positioning.

Figure 4.2.5   Quadcopter drone with the Raspberry PI Zero W board (A), a 500 mAh battery (B) and the high-resolution 3D positioning device (C) showing the set of four Vive sensors (D).

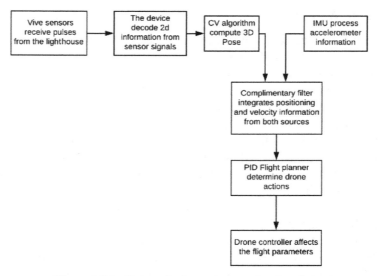

Figure 4.2.6   Data gathering and processing flowchart.

4.2.7 Conclusion

In this paper, we tested an approach to effectively perform 3D positioning in confined spaces at a high refresh rate, using a cheap device and an HTC Vive Lighthouse. We performed preliminary tests on flying devices to determine the feasibility and practicality and we showed initial results.

4.2.8 Designs and Source Code

Electronics design blueprints, printable 3D models and source code can be downloaded and shared freely from: https://github.com/germanespinosa/vive_device

References

[1] Gowda, Mahanth, Justin Manweiler, Ashutosh Dhekne, Romit Roy Choudhury, and Justin D. Weisz. "Tracking drone orientation with multiple GPS receivers." In Proceedings of the 22nd Annual International Conference on Mobile Computing and Networking, pp. 280–293. ACM, 2016.

[2] Mandal, Atri, Cristina V. Lopes, Tony Givargis, Amir Haghighat, Raja Jurdak, and Pierre Baldi. "Beep: 3D indoor positioning using audible sound." In Consumer Communications and Networking Conference, 2005. CCNC. 2005 Second IEEE, pp. 348–353. IEEE, 2005.

[3] Lim, Hyuk, Lu-Chuan Kung, Jennifer C. Hou, and Haiyun Luo. Zero-configuration, Robust Indoor Localization: Theory and experimentation. 2005.

[4] Sanchez-Lopez, Jose Luis, Jesus Pestana, Paloma de la Puente, Adrian Carrio, and Pascual Campoy. "Visual quadrotor swarm for the imav 2013 indoor competition." In ROBOT 2013: First Iberian Robotics Conference, pp. 55–63. Springer, Cham, 2014.

[5] Niehorster, Diederick C., Li Li, and Markus Lappe. "The accuracy and precision of position and orientation tracking in the HTC vive virtual reality system for scientific research." i-Perception 8, no. 3 (2017): 2041669517708205.

[6] Schweighofer, Gerald, and Axel Pinz. "Robust pose estimation from a planar target." IEEE Transactions on Pattern Analysis and Machine Intelligence 28, no. 12 (2006): 2024–2030.

[7] Zheng, Wei, Fan Zhou, and Zengfu Wang. "Vision Based Robust Pose Estimation System in Different Observed Situations for an Outdoor Quadrotor [J]." International Journal of Control & Automation 7, no. 7 (2014): 65–84.

[8] Argentim, Lucas M., Willian C. Rezende, Paulo E. Santos, and Renato A. Aguiar. "PID, LQR and LQR-PID on a quadcopter platform." In Informatics, Electronics & Vision (ICIEV), 2013 International Conference on, pp. 1–6. IEEE, 2013.

4.3 Augmented Reality Interaction vs. Tablet Computer Control as Intuitive Robot Programming Concept

Franz Steinmetz[1] and Annika Wollschläger[2]

[1]German Aerospace Center (DLR) Institute of Robotics
and Mechatronics, Germany
[2]Franka Emika GmbH, Germany

Intuitive programming concepts allow non-experts to undertake programming tasks. They are therefore of interest for small and medium-sized enterprises (SMEs) to reduce the programming effort required for automation. This work builds upon an integrated framework for task-level programming and task execution allowing a non-expert to program robots via a human-robot interface (HRI) and a graphical user interface (GUI) by defining and parameterizing simple program sequences. As an alternative input method to touch, a novel interaction concept based on augmented reality (AR) and haptic inputs are developed. The robot to be programmed serves as the input, i.e. one can push against its end-effector (e.g., from the left or right) to trigger certain actions. The available haptic input options, which are context-specific, are superimposed directly at the end-effector using AR glasses. We evaluate the usability of this interaction concept in comparison to a tablet computer control in a user study by means of subjective and objective data. The results indicate that the AR interaction concept can be a useful extension to touch interfaces in situations where interaction with the robot is required in either case.

4.3.1 Introduction

Industrial robots for manufacturing allow for an increase in productivity and a reduction in the price per unit compared to conventional production processes. However, the usage of such automation solutions is currently not applicable for small and medium-sized enterprises (SMEs), because conventional systems cause high programming costs and cannot be adapted easily to varying tasks. Due to just-in-time production, the increasing number of varieties and shortened product life cycles, the demand for flexible and applicable automation solutions with a high variety of use cases is present in large companies as well. The replacement of expert systems for robot programming by intuitive programming concepts can make the use of industrial robots interesting for SMEs because non-experts are put in the position to undertake programming tasks [13].

Such an integrated framework for task-level programming and task execution using robot skills is developed in [14] and [15] called RAZER. It allows a non-expert to program a robot via a human-robot interface (HRI). A graphical user interface (GUI) on a tablet computer is used to parameterize skills, sequence them to tasks and finally execute and monitor those tasks. Programming by Demonstration (PbD) is used to define poses and trajectories. The user is able to move the robot in gravity compensation mode guided by dialogs on the GUI to the desired positions.

As all information and instructions are provided by the GUI, problems arise if the tablet computer is not readily available. While teaching poses and trajectories, the user has to switch repeatedly between the operation of the tablet computer and the robot. Alternatively, the robot is moved to the desired position and the tablet computer is used for confirmation and continuation. As the tablet computer is put aside, the user has no opportunity to react to possible error messages appearing on the screen of the tablet computer.

Therefore, a new interaction concept is proposed. The robot to be programmed acts as a haptic input device. It registers pushes against its end-effector in different directions by analyzing the external forces working upon the end effector. The available input options are presented to the user via augmented reality (AR) glasses, directly superimposing the end-effector of the robot, comparable to a head-up display. Furthermore, (error) messages visible on the tablet computer are shown in the field of view (FOV) of the user. Additionally, head gestures (e.g., nodding) are registered, to enable the user to confirm a pose or cancel the teaching procedure in gravity compensation mode, when the robot cannot be used as input device. The interaction concept in this work is novel in that it provides a way of input and information provision, where the user does neither have to interrupt the direct physical interaction with the robot nor have to remember the available commands.

Our hypothesis is that while in general, touch-based interaction has a high usability, the AR interaction concept can be superior in situations, where the user is required to operate the robot in either case, for example for PbD. To test this hypothesis, we conducted a user study, comparing the AR interaction concept with touch-based interaction on a tablet computer.

The contribution of this work, therefore, includes (i) the concept of a novel AR-based interaction concept, (ii) the integration of this concept into our task-programming framework, and (iii) the comparison of the AR concept with established touch-based interaction on a tablet. This allows us to evaluate the suitability of each concept for different situations.

The remainder of this work is structured as follows. The related research is introduced in Section 4.3.2. After presenting the AR interaction concept in Section 4.3.3, the implementation is explained in detail in Section 4.3.4. The usability of the AR interaction concept was evaluated in a user study in comparison to the usability of a tablet computer control, which is described in Section 4.3.5. The results of this study are described in Section 4.3.6 and subsequently discussed in Section 4.3.7. The paper concludes with Section 4.3.8.

4.3.2 Related Work

Robot programming can be divided into manual and automatic programming [3]. For the former, the desired behavior of the robot is directly specified, usually using text-based offline programming. On the contrary, automatic robot programming permits only limited control of the code. PbD is one form of automatic programming. PbD describes the inference of a program by recording a sequence of actions from a user [4].

4.3.2.1 User interfaces

For the communication between humans and robots, interfaces are necessary to translate from human language to language understood by the robot. Human-robot interfaces (HRIs) can be distinguished in GUIs, tangible user interfaces (TUIs) and natural user interfaces (NUIs) [5].

GUIs let a system be operable via graphical symbols and control elements. They are usually displayed on screens and are operated via computer mice, joysticks or touch.

TUIs allow interactions by means of physical representations. TUIs could facilitate the programming process, especially for novice users, as suggested by the findings in [11].

NUIs use natural forms of interactions, e.g., touch, speech or gestures. It is shown in [17] that natural input methods are more easily understood by non-experts than conventional input methods.

4.3.2.2 Augmented reality in human-robot interaction

In this work, AR is used to design an interaction concept used in combination with a GUI and PbD. AR describes the superimposition of virtual elements with the real world by means of optical-see-through (OST) or video-see-trough (VST) devices. OST devices allow the user to see the physical environment directly while the virtual elements are displayed on a screen in his FOV, whereas VST devices visualize virtual elements in combination with the physical environment captured by a camera on a single screen.

There are several examples, where AR is used for HRI. In [1], the user can select objects for a pick-and-place task in an AR environment and receives visual feedback during the procedure. In [6], the user defines a collision-free workspace as well as the start and target configuration to compute the path of the robot. An OST device is used in [18], where coordinates for path planning are defined by the user for paintwork. A graphical representation of the paintwork is projected directly onto the target object via AR.

The combination of AR with NUIs can be found among others in [7], where the user can define poses and trajectories via gesture control. The capture of the gestures as well as the graphical display of the defined poses and trajectories is done in an AR environment with a VST device. In [9], a VSTdevice, as well as a 3D-pointing device, is used to define poses and paths around virtual objects. In [2], a visuohaptic AR system for programming tasks by demonstration is developed. A haptic user interface allows the user to perform manipulation actions within an AR environment.

The use of AR laser projection is described in [23]. Trajectories and target coordinates are projected into the direct environment of a robot.

In contrast to the described applications, the new interaction concept has no need for an additional input device like a handheld or pointing device but uses the robot to be programmed as an input device. It only requires AR glasses to be worn, which are equipped with a camera and an inertial measurement unit (IMU) for localization.

4.3.3 Interaction Concept

The AR interaction concept requires a robot to be equipped with some kind of force sensor to register the forces working upon the end-effector. Furthermore, it has to be suitable to work together with humans in direct physical contact.

4.3.3.1 Input directions

The robot serves as a haptic input device by registering forces exerted at its end-effector. In our concept, we focus on pressing the end-effector down, up, to the right, to the left or to the back, thus five different input directions can be differentiated. The concept could be extended to include for example rotations. Depending on the pose of the robot, those directions are not clearly defined. Two different input concepts were taken into account (see Figure 4.3.1):

(a) End-effector-centered input concept: The orientation of the end-effector determines the input direction. Up and down are registered by means of forces along the z-axis of the end-effector, left and right along the x-axis and back along the y-axis.

(b) Base-centered input concept: The input directions up and down are independent of the orientation of the end effector. Down is always a push in the direction of the gravity vector, up in the opposite direction. The push from the front points always in the direction of the base on a plane perpendicular to the gravity vector. Left and right are registered by pushes in the same plane and perpendicular to the vector determining pushes to the back.

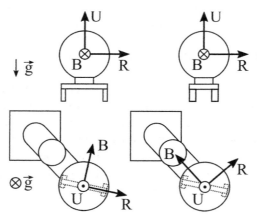

Figure 4.3.1 Two concepts of input directions: end-effector-centered (left two) and base-centered (right two). The directions are labeled with R (right), U (up) and B (back), the base is visualized as square, the end-effector as circle with gripper. The upper two views show into the back direction (B), note the rotated end-effector in the upper-right view. The lower two views follow the gravity vector (top-down).

4.3.3.2 Allocation of input directions

Depending on the state and context of the software, there are different input options available. The number of input options can vary from zero to several dozens. Those options have to be allocated to the five input directions of the new interaction concept, see Figure 4.3.2.

Hereby, often occurring input options are allocated always to the same input direction for intuitiveness. Close-related actions are always allocated to the input direction down, whereas input options like Next or Previous are allocated to the input directions right and left, respectively.

If there are more than five input options available at the same time, one input direction is used to allocate the next four input options from the list of all available input options.

4.3.3.3 Design of virtual elements

Virtual elements are needed to display the input options superimposed with the real world in the FOV of the user. Even though the design of 2D elements for GUIs is much more established than the design of 3D elements for AR interfaces, there exist some guidelines [8, 10, 20] for AR which could be condensed to:

- Present textual information clearly
- Ensure a sufficient contrast between virtual elements and the background

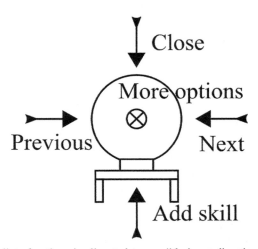

Figure 4.3.2 The list of options is allocated to possible input directions. Common options are always assigned to the same direction (such as "Close" to down). If there are more options than directions, one direction is used to change the allocation.

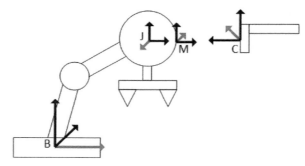

Figure 4.3.3   Base frame B, joint frame J, marker frame M attached to the robot (left) and camera frame C attached to the AR glasses (right). The x-axes are marked red.

- Group related elements
- Do not obscure important elements in the real by virtual elements
- Enable the user to identify information which needs attention
- Keep virtual elements simple

4.3.3.4 Display orientation and position

To display the virtual elements with the correct pose into the view of the user, the transformation from the camera within the AR glasses (frame C) to the center of the last joint of the robot arm (frame J) $^{C}_{J}T_{BC}$ has to be determined. Figure 4.3.3 shows the frames attached to the robot and the AR glasses.

4.3.3.5 Push detection

F_J is the force relative to the center of the last robot joint. If F_J exceeds a certain threshold value in any direction, the interaction is considered as a valid input and the corresponding input option is triggered. To consider another interaction as a valid input after the first one, the force has to drop temporarily below the threshold value.

4.3.3.6 Gesture detection

If the robot is in gravity compensation mode during PbD, the described input method can no longer be used, because external forces lead to a movement of the robot. The user typically has two input options available during gravity compensation mode, i.e., to cancel a (teaching) processor to confirm (a demonstrated pose or trajectory). Therefore, the IMU of the AR glasses is used to detect head shakes for cancellation and head nods for confirmation.

Figure 4.3.4   Using the proposed AR interaction concept, the operator uses the robot end-effector as haptic input device. Context-specific input options are superimposed using the AR glasses.

4.3.4 System Implementation

The new interaction concept is integrated into RAZER [14, 15]. It consists of a web service that manages the communication between the frontend of RAZER, the AR glasses, and the robot. In this work, seven degrees of freedom KUKA lightweight robot (LWR) is used as a robot and an Epson Moverio BT-200 as AR glasses, see Figure 4.3.4. The application for the glasses on the Android device is based on the Vuforia software platform.

4.3.4.1 Input directions
As indicated by a first survey, most people seem to find the base centered input concept more intuitive, which is why it is implemented for the evaluation of the AR interaction concept.

4.3.4.2 Allocation of input directions
When the robot programming is started, an object is created, which contains a list of all available input options. Input options are for example Open new task, Add skill or Set velocity to 0.6 [m/s]. The object is updated, whenever the available input options change. This happens, when the RAZER frontend changes its state.

4.3.4.3 Design of virtual elements

Different designs of the virtual elements were considered under the aspects of intuitiveness and visibility. The design of the virtual elements is challenging due to the small FOV, the resolution and the brightness of present AR glasses. Figure 4.3.5 shows the final design of the virtual elements for five input options. To achieve maximal contrast between the real world and the virtual elements, latter is displayed in white.

All input options are displayed by means of a simple geometric shape with the icon representing the action in the center. The geometric shapes are circles (for actions), diamonds (for groups) and squares (for navigational actions such as back or more) respectively. The shapes are flanked by a 3D hand indicating the pushing from the left, right, top, bottom or front. Shape and hand are complemented by the name of the input option. Related elements are grouped together. The black border of the textual information will appear on the OST devices transparent. It is used to make the textual information readable if it overlaps other virtual elements.

Figure 4.3.6 shows the display of an error message which can be confirmed by the user by pushing from the left.

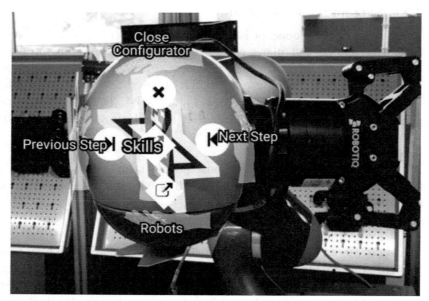

Figure 4.3.5 Virtual elements are superimposed at the end-effector into the FOV of the user. Here, five input options are possible.

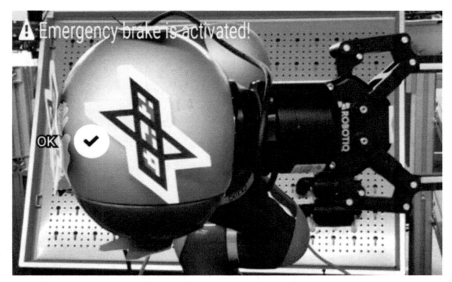

Figure 4.3.6 Also messages are displayed using the AR glasses. The image shows an error message, which can be confirmed by the user.

4.3.4.4 Display orientation and position

Several markers (frames Mi) are attached to the robot joint. These AR tags with the shape of the German Aerospace Center (DLR) logo can be well seen in Figure 4.3.4. They have a known orientation and distance to the robot joint's center point, such that camera-based tracking can be used to determine $^{C}_{M_i}T$. The transformation from the marker to the center point of the robot $^{M_i}_{J}T_{BC}$ depends on the orientation and position of the marker as well as the pose of the robot from the base (frame B) to the end-effector (frame E) $^{B}_{E}T$, to take into account the base-centered input concept. The pose of the robot, as well as its joint configuration, can directly be readout. $^{M_i}_{J}T_{BC}$ is determined via

$$^{M_i}_{J}T_{BC} = \begin{pmatrix} {}^{B}_{J}R^{-1} \cdot R_{BC} & \begin{matrix} 0 \\ 0 \\ 0 \end{matrix} \\ 0 \quad 0 \quad 0 & 1 \end{pmatrix} \cdot {}^{M_i}_{J}T, \qquad (4.3.1)$$

where $^{B}_{J}R$ is the transformation between the base and the joint of the robot and $^{M_i}_{J}T$ is a constant for each marker M_i, taking into account its orientation and the radius of the robot joint. R_{BC} is a rotation of α_{BC} about the z-axis of the base frame. A_{BC} is determined by means of the position of the joint relative to the base of the robot $_J^B p$ via

$$\alpha_{BC} = \arctan\left(\frac{_J^B p_y}{_J^B p_x}\right) \qquad (4.3.2)$$

for $_J^B p_x 6 = 0$. R_{BC} can then be determined via

$$R_{BC} = \begin{pmatrix} \cos\alpha_{BC} & -\sin\alpha_{BC} & 0 \\ \sin\alpha_{BC} & -\cos\alpha_{BC} & 0 \\ 0 & 0 & 1 \end{pmatrix} \qquad (4.3.3)$$

4.3.4.5 Push detection

The external forces F_{ext} acting upon the end-effector of the robot are continuously read out. The force F_J relative to the robot joint can be determined via

$$F_J = R_{BC-1} \cdot {_{BE}R_{-1}} \cdot F_{ext}, \qquad (4.3.4)$$

where $_E^B R$ is the rotation between base and end-effector frame of the robot.

4.3.4.6 Gesture detection

The switch between both modes is indicated by means of a message in the FOV of the user giving instructions on the available input gestures (see Figure 4.3.6 for a message in the FOV).

4.3.5 Evaluation

The usability of the AR interaction concept and the tablet computer control was evaluated in a user study. A standardized questionnaire (QUEAD [12] was used to assess the subjective rating of usability. The efficiency of the approaches was measured by means of time to completion and operating errors of a test task. The test task was first performed using solely the GUI on the tablet computer and afterward with the AR interaction concept. The usability questionnaire was answered directly after solving the task with each system respectively.

Seven test persons participated in the study. All were students working at the German Aerospace Center (DLR). Five were male, two were female. They were on average 24.1 ± 1.9 years old. All were right-handed and needed glasses for visual aid. Two had never used an AR device before, everybody else rarely. All test persons were regularly using touch screen devices.

The study was conducted as follows. After answering a questionnaire collecting demographic data and past experiences, the AR glasses were calibrated for each test person to take into account individual facial geometry. To get the test person used to the systems, step-by-step instructions for a task were provided, which the test person had to follow for each of the systems before starting the actual test task.

A test task was defined, in which the test person had to program a sequence consisting of two skills. Skill 1 involved moving the robot (see Figure 4.3.4) to a defined pose with a velocity of 0.6 [m/s]. Skill 2 was a Pick & Place task, in which the robot should be instructed to move an object from position A to position B. During execution of the test task with the AR interaction concept, the test person was asked to use the GUI as little as possible.

4.3.6 Results

Table 4.3.1 and Figure 4.3.7 shows the results of the usability questionnaires. The usability was rated in the categories Perceived usefulness, Perceived ease of use, Emotions, Attitude, and Comfort. Perceived usefulness describes the test person's subjective perception of the ability of a system to increase the performance of a given task. The perceived ease of use is the degree to which the test person perceives the system's use as free from physical and mental strain. Attitude toward using technology is defined as the test person's overall affective reaction to using the system [19].

Table 4.3.2 and Figure 4.3.8 show the results for the time to completion. The following operation errors were made while using the AR interaction concept:

- Two of the test persons did not spend enough force on the robot to trigger an input
- After the termination of gravity compensation mode, three test persons tried to confirm a dialog with a head nod

Table 4.3.1 Mean and standard deviation of the ratings of the usability questionnaire from 1 (very poor) to 5 (very good)

	GUI	AR
Perceived usefulness	4.33 ± 0.50	3.10 ± 0.91
Perceived ease of use	4.41 ± 0.22	3.27 ± 0.92
Emotions	4.71 ± 0.28	4.05 ± 0.73
Attitude	4.57 ± 0.32	3.76 ± 1.07
Comfort	4.76 ± 0.50	4.19 ± 0.88

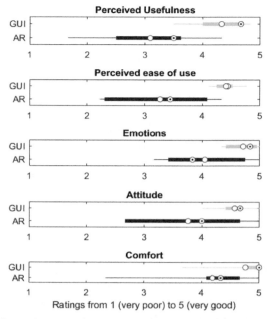

Figure 4.3.7   Ratings of the usability questionnaire with mean (∘), median (⊙), upper and lower quartile (■) and range (−).

Table 4.3.2   Mean and standard deviation of time to completion in minutes

	GUI	AR
Overall time [min]	3.18 ± 1.24	6.42 ± 1.42
Time PbD [min]	1.22 ± 0.53	1.30 ± 0.55

- Two of the test persons tried to push the robot during a gravity compensation mode to trigger an input
- One test person moved his head during gravity compensation mode, such that an unintended head nod was registered

4.3.7 Discussion

4.3.7.1 Objective data

The efficiency was evaluated by means of time to completion and user errors. When looking at the overall time to completion, users took about twice as long for the AR interaction concept compared to the tablet interface. The users were prompted to use the GUI as little as possible during use

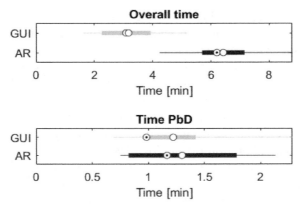

Figure 4.3.8 Completion times with mean (○), median (⊙), upper and lower quartile (■) and range (−).

of the AR interaction concept. Therefore, the parametrization of the skills, where the user has a huge number of input options, was accomplished solely by interacting with the robot. However, the interaction concept provides a maximum of five options at once, thus a lot of time is needed to navigate to the desired option. The interaction paradigms of GUIs with sliders, text boxes or drop-down menus are therefore oftentimes superior to the AR interaction concept, especially when there are more input options than interaction options. Focusing only on the time to completion for subtasks involving PbD, the durations are comparable for the two concepts. In this subtask, the user had to interact with the robot in either case and the number of input options was smaller than five, which is both advantageous for the AR interaction concept.

As mentioned above, the test persons made several user errors while using the AR interaction concept. The users were able to recognize that their approach did not lead to the desired results and were nonetheless able to successfully complete the task. Only in the case where an unintended head nod was detected, a repetition of the PbD procedure was necessary to complete the task.

4.3.7.2 Subjective data

The perceived usefulness of the AR interaction concept is rated clearly worse than for the GUI on the tablet computer. This can be traced back to the low efficiency of the system for tasks with a lot of input options (see Section 4.3.7.1). The observations match the findings in [21], where the performance in an assembly task using a tablet computer and a head-mounted display was compared. Results showed a higher performance while working with the tablet computer. The small FOV of the AR glasses required to display the virtual elements in minimum space. This causes an overlapping of icons and text elements, which possibly led to a worse rating of the perceived ease of use for the AR interaction concept. Another influence on the perceived ease of use could be the perception of double images, which was reported by two test persons. The occurrence of double images can be traced back to an inadequate calibration of the AR glasses. All test persons regularly use touch screens and are therefore familiar with the interaction paradigms of GUIs, whereas AR devices are never or seldom used by the test persons. Even though the test users executed simple tasks with both interaction concepts before the actual test task, they had to get used to the new technology during the study. In [21], test users experienced almost no effects of familiarizing after four hours of using a head-mounted display. The findings suggest that there is much more time needed to get used to such a new system, which could have an effect on the perceived ease of use.

Regarding emotions and attitude, AR is also rated worse than the GUI, yet still positive. In [16], it is shown, that well-designed systems and a smaller degree of novelty lead to more positive emotions. Again, a more frequent use of AR devices could, therefore, lead to more positive emotions and a better attitude.

All test persons needed glasses for vision aid, which have to be worn in addition to the AR glasses. One of the test persons reported pain during wearing the AR glasses. In [22], it is reported that a head-mounted display led to head and neck pain after prolonged use, whereas lighter and more compact smart glasses did not. The rating of the comfort of the AR interaction concept could, therefore, be further improved by using lighter and more compact glasses.

4.3.8 Conclusion

In this paper, we introduced a novel interaction concept using the robot as haptic input device together with AR to display context-specific input options. The conducted user study confirmed our hypothesis about its usability for certain situations. Tablet computer control is generally fast and intuitive. However, when the user is required to teaching poses or trajectories using the robot, the tablet is not within reach. Here, the results of the study suggest that the AR interaction concept can be advantageous. Still, for the maturation of the concept, several improvements are required. Especially the used AR glasses have hardware restrictions limiting the current usability, but also the overall robustness can further be improved. The AR interaction concept cannot serve as a replacement to common touch interfaces but promises to be a useful extension.

For the future, the input options using a robot can further be explored. On the one hand, the concepts of end-effector centered, and base-centered inputs need to be investigated further. On the other hand, more input options should be considered, to e.g., allow for rotations or pushes at the elbow.

References

[1] B. Akan, A. Afshin, B. Curuklu, and L. Asplund, Intuitive industrial robot programming through incremental multimodal language and augmented reality, in IEEE Int. Conf. Robotics and Automation, pp. 3934–3939, 2011.

[2] J. Aleotti, G. Micconi, and S. Caselli. Asplund, Object interaction and task programming by demonstration in visuo-haptic augmented reality, in Multimedia Systems, vol. 22, pp. 675–691, 2016.

[3] G. Biggs, and B. MacDonald, A survey of robot programming systems, In Proc. Australasian Conf. Robotics and Automation, pp. 1–3, 2003.

[4] A. Billard, S. Calinon, R. Dillmann, and S. Schaal, Robot Programming by Demonstration, In Springer Handbook of Robotics, pp. 1371–1394, 2008.

[5] I. I. Bittencourt, M. C. Baranauskas, R. Pereira, D. Dermeval, S. Isotani, and P. Jaques, A systematic review on multi-device inclusive environments, In Universal Access in the Information Society, vol. 15, pp. 737–772, 2016.

[6] J. W. S. Chong, S. K. Ong, A. Y. C. Nee, and K. Youcef-Youmi, Robot programming using augmented reality: An interactive method for planning collision-free paths, in Robotics and Computer-Integrated Manufacturing, pp. 689–701, 2009.

[7] J. Lambrecht, and J. Kruger, Spatial programming for industrial robots based on gestures and augmented reality, in IEEE Int. Conf. Intelligent Robots and Systems, pp. 466–472, 2012.

[8] S. Ganapathy, G. J. Anderson, and D. K. Marsh, Techniques for mobile augmented reality applications, U.S. Patent No. 9,264,515, 2016.

[9] A. Gaschler, M. Springer, M. Rickert, and A. Knoll, Intuitive robot tasks with augmented reality and virtual obstacles, in IEEE Int. Conf. Robotics and Automation, pp. 6026–6031, 2014.

[10] C. Rolim, D. Schmalstieg, D. Kalkofen, and V. Teichrieb, Design Guidelines for Generating Augmented Reality Instructions, in IEEE Int. Conf. Mixed and Augmented Reality, pp. 120–123, 2015.

[11] T. Sapounides, and S. Demetriadis, Tangible versus graphical user interfaces for robot programming: exploring cross-age childrens preferences, in Personal and Ubiquitous Computing, vol. 17, pp. 1775–1786, 2013.

[12] J. Schmidtler, K. Bengler, F. Dimear, and A. Campeau-Lecours, A Questionnaire for the Evaluation of Physical Assistive Devices (QUEAD), in IEEE Int. Conf. Systems, Man, and Cybernetics, 2017.

[13] R. D. Schraft, and C. Meyer, The need for an intuitive teaching method for small and medium enterprises, VDI Berichte, vol. 1956, p. 95, 2006.

[14] F. Steinmetz, and R. Weitschat, Skill Parametrization Approaches and Skill Architecture for Human-Robot Interaction, in IEEE Int. Conf. Automation Science and Engineering, pp. 280–285, 2016.

[15] F. Steinmetz, A. Wollschlager, and R. Weitschat, Human-robot interface with integrated framework for visual task-level programming and intuitive parameterization, submitted to IEEE Int. Conf. Robotics and Automation, 2018.

[16] M. Thuring, and S. Mahlke, Usability, aesthetics, and emotions in human-technology interaction, in Int. Journal of Psychology, vol. 42, pp. 253–264, 2007.

[17] P. Tsarouchi, A. Athanasatos, S. Makris, X. Chatzigeorgiou, and G. Chryssolouris, High-Level Robot Programming Using Body and Hand Gestures, in Procedia CIRP, vol. 55, pp. 1–5, 2016.

[18] T. Pettersen, J. Pretlove, C. Skourup, T. Torbjorn, and T. Lkstad, Augmented reality for programming industrial robots, in IEEE Int. Conf. Mixed and Augmented Reality, p. 319, 2003.

[19] V. Venkatesh, M. G. Morris, G. B. Davis, and F. D. Davis, User acceptance of information technology: Toward a unified view, in MIS quarterly, vol. 27, pp. 425–478, 2003.

[20] R. Wetzel, R. McCall, and A.-K. Braun, W. Broll, Guidelines for designing augmented reality games, in Proc. Conf. Future Play, pp. 173–180, 2008.

[21] M. Wille, B. Grauel, and L. Adolph, Strain caused by head-mounted displays, in Proc. Human Factors and Ergonomics Society Europe, pp. 267–277, 2013.

[22] M. Wille, P. M. Scholl, S. Wischniewski, and K. Van Learhoven, Comparing Google glass with tablet-PC as guidance system for assembling tasks, in Wearable and Implantable Body Sensor Networks Workshops, pp. 38–41, 2014.

[23] M. F. Zaeh, and W. Vogl, Interactive laser-projection for programming industrial robots, in IEEE Int. Conf. Mixed and Augmented Reality, pp. 125–128, 2006.

4.4 A Multimodal System Using Augmented Reality, Gestures, and Tactile Feedback for Robot Trajectory Programming and Execution

Wesley P. Chan[1,*], Camilo Perez Quintero[1,*], Matthew K. X. J. Pan[1], Maram Sakr[1], H. F. Machiel Van der Loos[1] and Elizabeth Croft[2]

[1]Collaborative Advanced Robotics and Intelligent Systems (CARIS) Laboratory, Mechanical Engineering University of British Columbia, Canada
[2]Monash University, Australia
*These authors contributed equally.

Existing industrial robot program interfaces, e.g., teach pendants and computer consoles, are often unintuitive, resulting in a slow and tedious teaching process. While kinesthetic teaching provides an alternative for small robots, where safe interaction can be guaranteed, for large industrial robots, physical interaction is not an option. Emerging augmented reality (AR) technology offers an alternative with the potential for faster, safer, and more intuitive robot programming as it admits the presence of rich amounts of visual, in-situ information. However, too much information may also overload the user's visual perception capacity, and it may not provide adequate feedback of the robot state. We present a multimodal system for trajectory programming and on-line control, merging AR, electromyography reading, gesture control, speech control, and tactile feedback. We explore how hidden task variables such as force can be presented and controlled by the user during execution. Preliminary evaluation of our system where the user specifies *in-situ* trajectories with various force profiles showed that tactile feedback is favored over visual feedback or the combination of both.

4.4.1 Introduction

For decades, teach pendants, augmented with computer consoles, has been the *de facto* interface for programming industrial robots. Over time, this programming modality has seen little change, mainly due to the infrequency of robot programming for fully automated tasks once an assembly operation is set-up. However, the recent introduction of less expensive and increasingly interactive robots has allowed for more flexibility in the manufacturing process. Thus, the infrequency of programming and reprogramming expected for traditional robots may not apply. As an example, kinesthetic teaching of robots such as Baxter, Sawyer, and the KUKA iiwa allows for easy and frequent reprogramming, permitting these robots to execute operations for customized products in small lot sizes, thus, enabling automated processing for a highly variable product mix. This capacity is part of an overall paradigm shift were industries that traditionally utilized siloed robotic manufacturing lines are opening up to the utilization of smaller scale reconfigurable robots and human-robot hybrid collaborative teams to handle complex production and assembly tasks. This shift also opens up opportunities for smaller producers with complex operations and/or small lot sizes [1–3].

With this shift comes new requirements for flexible and intuitive methods to reprogram and interact with such robots, in order to ensure safety and efficiency. Along with kinesthetic teaching methods, emerging AR technology provides a promising alternative to traditional teach pendants for addressing such requirements. With increasing complexity of industrial robotic systems, which may not be safe for physical interaction, there is a growing demand for alternative robot programming user interfaces. The alternative should provide sufficient capacity for communicating all necessary information to the user, without adding a layer of complexity that distracts the user from the task [4]. Traditional programming methods lack such capacity and often result in a cumbersome interaction. AR enables the creation of a rich set of user interfaces that are co-located with the robot, allowing the user to have better situational awareness [5]. Furthermore, it permits visualization and interaction with hidden process variables that are not exposed to the operator in traditional programming methods (e.g., force, velocity, acceleration). By improving the quality of shared information between the human and robot, we can achieve more effective human-robot interaction [6].

With AR devices and development tools such as the Microsoft HoloLens [7], Epson Moverio [8], and Magic Leap [9] increasingly available, researchers have explored the use of augmented reality for various tasks including assembly [10], maintenance [11], repair [12], and training [13] and found positive results. While augmented reality can provide a rich amount of visual information, too much information can cause perceptual overload [14]. Multimodal systems have been suggested to be more efficient due to their similarity with daily interactions [15]. Furthermore, not all task-relevant information is best communicated visually. For example, higher-order motions (e.g., accelerations and jerks), as well as forces and moments (e.g., friction, torsion) can be difficult to visualize. In collaborative tasks, humans often utilize multiple inputs and communication channels including haptics, gestures, and gazes, and studies have shown that using a combination of communication channels for human-robot cooperation can achieve more effective interaction [16].

In this paper, we extend our recent work in robot programming [17], by developing a visual-tactile interface utilizing electromyography signals and vibrotactile feedback. The user specifies a virtual path on a physical surface and then controls the robot's end-effector velocity through the path and the amount of exerted force. We provide a preliminary demonstration of our interface in which a user is asked to specify a path and follow three different virtual force profiles.

4.4.2 Related Work

4.4.2.1 Augmented reality

AR technology has shown potential for improving human-robot interactions [18] and robot programming [19]. Zaeh et al. [20] proposed an industrial robot programming interface. Trajectories and target coordinates are projected in AR onto the robot's environment and can be manipulated interactively. A laser projection technique allowed the user to intuitively draw desired motion path directly on the workpiece. Chong et al. [21] introduced a methodology for planning collision-free paths in AR environment. In addition, they proposed the usage of a scalable virtual robot that offers flexibility and adaptability to different environments when an in-situ robot programming approach is desired. Their system allowed the user to define the start, intermediate, and end goal configurations, and preview a simulated path before executing on the real robot. Green et al. [22] proposed an AR teleoperating system for mobile robots and compared them with traditional systems using images from a robot-mounted camera for feedback. Their AR system provides an exocentric view of the robot and allows for gesture and speech interaction. They showed that the AR system yields better task completion accuracy and situational awareness.

4.4.2.2 Force feedback

While the uptake for AR-based robot programming systems is growing, most current AR systems lack a force feedback channel. However, force feedback during task execution is important and informative for achieving task success [23–25]. Force feedback can be provided visually through a virtual gauge or meter [26], or through an in-hand haptic device [27]. Using a visual gauge may be unintuitive since it maps a haptic sense to visual input, while using an in-hand haptic device requires the user to hold on to the device, removing the opportunity for the user to complete other work and, if the device is large or tethered, preventing the user from moving around the workspace to observe the work.

4.4.2.3 Haptics

Tactile displays offer users an unobtrusive method of receiving information and has been found to be particularly useful during mentally demanding contexts. Using haptic feedback with, or in lieu of, visual displays have been shown to reduce overall mental workload relative to using only visual display of information [28]. Several studies show that tactile information elicits faster reaction time and reduces mental effort [29].

Much of the early research in tactile displays have been oriented towards assisting users in navigating real and virtual environments [29]. However, this effort has since expanded towards providing cues related to timing [30] or even user psycho-physiological state [31]. With the rise of wearables and smartwatches within the past few years, the ubiquity of wearable vibrotactile interfaces has formatively transitioned wearable haptics from a niche research area into a commonplace feature for smart devices. Within the context of this work, our goal is to integrate haptics in an AR interaction system as a low-attention feedback mechanism to allow users more control over a human-robot interaction.

4.4.2.4 Gestures

Gestures are widely used by humans as a way of communication, and many researches have found applications to human-robot interaction. Yu et al. created a system allowing users to carry out the complex task of controlling multiple drones using gestures along with position and color information [32]. Pan et al. explored gestures in the context of human-robot handover and found certain cues exist that are indicative of an impending handover [31]. Similarly, Moon et al. demonstrated that human hesitation gestures can be applied to robots for achieving more effective human-robot collaboration. Haddadi et al., also identified and demonstrated human arm gestures that can be implemented onto a robotic arm for conveying robot intent in a collaborative manufacturing setting [33]. Hence, gesture can be used for effective communication between humans and robots.

To allow users to intuitively program robotic trajectories and execute force-relevant tasks successfully, we propose a hands-free, untethered system combining the use of augmented reality with gesture control and tactile and visual force feedback. Specifically, in this paper, we propose using multi-modal control and feedback to avoid overloading single communication channels. Many robotic tasks require the robot to move along a specified trajectory while applying a specific force profile along the path. Grinding, polishing, and welding are examples of such tasks. Here, we evaluate the usage of visual and haptic feedback to help the user maintain the prede-fined force profile while controlling the robot to move along a specific trajectory.

4.4.3 System

4.4.3.1 Hardware

Our system is comprised of two main interaction platforms: the Microsoft HoloLens, the first untethered head-mounted AR system [7], and the Thalmi-cLabs Myo armband the first portable gesture-based input device using sEMG [34]. The HoloLens allows us to render 3D virtual objects in the physical world and provides speech recognition for audio input from the user. It can also track the user's gaze. The Myo armband provides muscle activation levels read from eight sets of electrodes. It also provides recognition of six simple hand gestures as well as the arm's acceleration and orientation. The Myo armband has a vibrator embedded which can be used to provide tactile feedback. In our current work, we use our system to control a 7 DOF Barrett Whole Arm Manipulator (WAM) (Figure 4.4.1).

4.4.3.2 Software

We use the Microsoft HoloLens SDK and Unity for developing our aug-mented reality application. Robot Operating System (ROS) is used for con-trolling and interfacing with the WAM and the Myo armbands. The software package rosbridge is used for communication between the HoloLens and ROS controlling the WAM.

Our software system allows the user to visualize the robot, create trajecto-ries using gaze and speech, preview the trajectory, execute the trajectory using gestures, control the trajectory's force profile online using muscle activation, and provides tactile and visual force feedback during execution. We describe each of these components below.

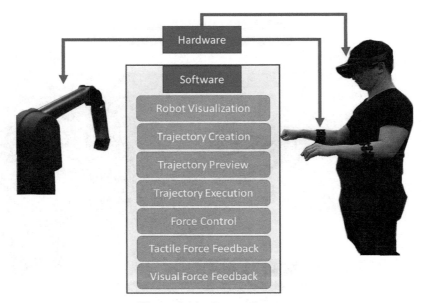

Figure 4.4.1   System diagram.

Robot visualization: Our system creates a kinematic model of the robot and renders a virtual robot on top of the real robot in physical space. As the person moves the robot in gravity compensation mode, the joint angles of the real robot are sent to the virtual robot. The virtual robot moves accordingly to match those of the real robot, providing an accurate visualization of the real robot (Figure 4.4.2).

Trajectory creation: The user's gaze is tracked by tracking the head orientation. A ray is traced out in the gaze direction and a red circular visual marker is rendered at its intersection with the environment. The user creates a trajectory by looking at key points on the surface in front of the robot and giving the speech command "set point" to set trajectory waypoints (Figure 4.4.3).

Trajectory preview: After each "set point" command is given, a green spherical marker with a blue normal arrow is created and rendered onto the surface to provide a visualization of the trajectory waypoints. When the user gives a verbal "lock path" command, a trajectory is generated using the waypoints $x_1, \ldots, x_n$ with a 3rd degree B-splines [35] in the form of:

$$B_i^k(x) = v_i^k(x)B_i^{k-1}(x) + \left(1 - v_{i+1}^k(x)\right) B_{i+1}^{k-1}(x) \qquad (4.4.1)$$

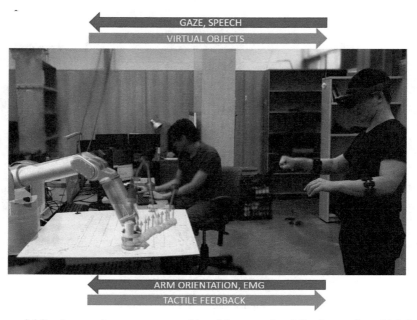

Figure 4.4.2 A user showcases our multimodal system by following a sinusoidal force pattern (pink line). The user controls the normal force exerted on the surface (blue arrow) and the end effector linear velocity by moving his forearm and changing his muscle activation level.

where,

$$v_i^k(x) = \frac{x - x_i}{x_{i+k} - x_i} \quad \text{and} \quad B_i^0(x) = \begin{cases} 1 & \text{if } x_i \le x < x_{i+1} \\ 0 & \text{otherwise} \end{cases}$$

Trajectory execution: During execution of the trajectory, the robot arm is constrained to the path using a force controller previously developed [36]. Given the end effector's current location x and the closest point on the path x_d, a restoring force F_s is applied in the direction of $\hat{s} = \hat{n} \times \hat{t}$, where $\hat{n}$ is the normal unit vector and $\hat{t}$ is the tangential unit vector of the path at point x_d. F_s is calculated as

$$F_s = K_{p_p}((x - x_d) \cdot \hat{s}) + K_{d_p}(\dot{x} \cdot \hat{s}) \tag{4.4.2}$$

where (K_{pp}) and (K_{dp}) are the proportional and derivative gains.

The user controls the robot's movement along the path using the Myo armband by moving the forearm left and right. The user begins trajectory

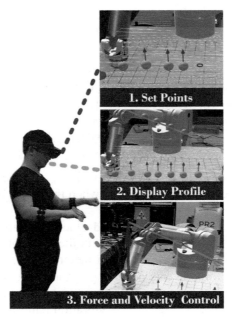

Figure 4.4.3 During our pilot study the user: sets the desired path using both head orientation and speech; displays the path profile using speech; and controls the robot arm force and velocity with a pair of MYO bands.

control by holding the right arm out with the forearm parallel to the ground and making a fist gesture. When the first gesture is first detected, the yaw orientation of the forearm θ_i is recorded. As the user moves the forearm, the yaw displacement is measured as $d\theta = \theta - \theta_i$, where θ is the current yaw angle measurement. A driving velocity proportional to $d\theta$ is then applied in the $\widehat{t}$ direction to move the end effector along the path.

Force control: The user controls the applied force along the $\widehat{n}$ direction by varying the muscle activation level. When the system initializes, the user first squeezes their right hand into a fist with maximum force, and a maximum EMG reading $\bar{\varepsilon}_{max}$ is recorded. The average EMG activation level at time t is computed as:

$$\varepsilon_{avg}(t) = \sum_{i=1}^{8} \varepsilon_i(t) \qquad (4.4.3)$$

where $\varepsilon_i(t)$ is the EMG reading of electrode i on the MYO band at time t. A smoothed average EMG signal is obtained by averaging $\varepsilon_i(t)$ over a time window:

$$\bar{\varepsilon}_{avg}(t) = \frac{\sum_i \varepsilon_{avg}(t-i)}{n} \tag{4.4.4}$$

The maximum EMG reading $\bar{\varepsilon}_{max}$ is taken as the $\bar{\varepsilon}_{avg}(t)$ value measured three seconds after the user was asked to hold a fist with maximum force to avoid initial spikes in the EMG reading. The force command is computed as:

$$F_n = F_{limit}\frac{\bar{\varepsilon}_{avg}(t) - \bar{\varepsilon}_{start}}{\bar{\varepsilon}_{max} - \bar{\varepsilon}_{start}} \tag{4.4.5}$$

where F_{limit} is set to 30 N, and $\bar{\varepsilon}_{start}$ is the $\bar{\varepsilon}_{avg}$ value measured at the time when trajectory execution began (i.e., when the MYO band first detected the fist gesture).

Tactile force feedback: We created a tactile pattern feedback module for providing force feedback to the user via the vibrators on the MYO bands. The module allows arbitrary vibration patterns to be sent to the user's arm to signal various force events occurring at the robot end effector. Given the target force F_{target} at a trajectory point, and a tolerance δ, if the applied force $F_{applied}$ is lower than $F_{target} - \delta$, a vibration pattern of quick, short pulses with one-second rest intervals is given. If $F_{applied}$ is higher than $F_{target} + \delta$, then a different pattern of longer pulses with one-second rest intervals is given. In our study, δ is set to 10 N.

Visual force feedback: To be able to compare different feedback mechanisms, we also enabled our system to provide visual force feedback. When the end effector applies a force to the environment, arrows with length proportional to the applied force magnitudes are rendered at the end effector to provide a visualization of the forces in the xyz axes.

4.4.4 Pilot

To demonstrate our system's usability, we piloted an experiment where a participant used the system to program and execute a trajectory with a given force profile. We tested three force profiles with different complexities: constant, ramp, and sinusoidal (Figure 4.4.4). For each of the force profiles, we also tested three modes of force feedback: tactile pattern, visual display, and tactile pattern with visual display.

In the pilot, the robot arm is located in front of a table with grid lines drawn on it, and the user stands across the robot. In each trial, the user first creates a 2D spatial trajectory on the table by looking at key points on the table and speaking the command word "set point". After the key

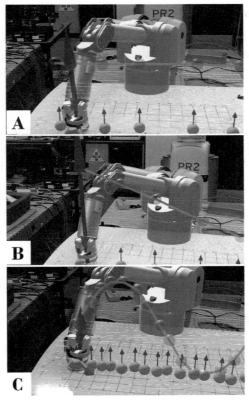

Figure 4.4.4 User's point of view. The pink line indicates force to follow: A. constant force profile. B. Ramp force profile, C. Sinusoidal force profile.

points have been set, the user then says "lock path" to generate a b-spline path through the key points and locking the robot's end-effector to the generated path, the robot is now constraint to the path but is free to move in the path direction. The user then says "display force", and the desired force profile is displayed above the 2D trajectory. When executing each trajectory, the user is asked to first keep the robot arm at the starting point of the trajectory and attain the target force level. Once target force level at the starting point has been achieved, the user then proceeds to move the robot along the trajectory, while trying to follow the force profile. Once the robot has reached the endpoint of the trajectory, the trial is complete (Figure 4.4.3).

4.4.5 Analysis

For each trial, we computed the execution time, t_{exec}, the maximum absolute force error, $|e|_{max}$, the average absolute force error, $|e|_{avg}$, and the cumulative absolute force error, $I_{|e|}$. The execution time is defined as

$$t_{exec} = t_{end} - t_{start},\qquad(4.4.6)$$

where t_{start} is the time when the robot first moves away from the starting point of the trajectory, and t_{end} is the time when the robot first moves past the endpoint of the trajectory. The maximum absolute force error is defined as

$$|e|_{max} = \max\{|e|(t)|t \in [t_{start}, t_{end}]\},\qquad(4.4.7)$$

$$|e|(t) = |F_{applied}(t) - F_{target}(t)|,\qquad(4.4.8)$$

while the average absolute force error is defined as

$$|e|_{avg} = \frac{\sum_{[t_{start},t_{end}]} |e|(t)}{n},\qquad(4.4.9)$$

and the cumulative absolute force error is defined as

$$I_{|e|} = \sum_{[t_{start},t_{end}]} |e|(t) * \Delta t,\qquad(4.4.10)$$

4.4.6 Result

The execution time, t_{exec}, maximum absolute force error, $|e|_{max}$, average absolute force error, $|e|_{avg}$, and cumulative absolute force error, $I_{|e|}$, measured in each trial, and the averages, are shown in Tables 4.4.1–4.4.4 respectively. Preliminary results show that in all conditions, the user was able to complete the trajectory within 17 s. $|e|_{avg}$ was within the tolerance $\delta = 10$ N for all trials, and $|e|_{max}$ did not exceed 20 N. The tactile case has the smallest average $|e|_{max}$, $|e|_{avg}$, and $I_{|e|}$ among all cases, while the visual case has the smallest average texec, with the t_{exec} measured in the tactile case only 0.1 s longer. Second, to the tactile case, the visual case has the next smallest $I_{|e|}$, performing better than the tactile and visual case, while the tactile and visual case has the next smallest $|e|_{max}$ and $|e|_{avg}$, performing better than the visual case. Figure 4.4.5 shows the plots of position and force error measured during the pilot, using the trial with tactile and visual feedback executing a sinusoidal force profile as an example.

Table 4.4.1 Execution time, t_{exec}, measured in each trial

t_{exec}(s)	Line	Ramp	Sin	Avg
Tactile	10.5	10.5	10.7	10.5
Visual	9.6	12.3	9.5	10.4
Tactile + Visual	11.3	13.2	16.9	13.8

Table 4.4.2 Maximum absolute force error, $|e|_{max}$, measured in each trial

$|e|_{max}$(N)	Line	Ramp	Sin	Avg
Tactile	16.2	6.1	10.6	11.0
Visual	12.0	19.8	17.8	13.5
Tactile + Visual	10.3	12.0	15.5	12.6

Table 4.4.3 Average absolute force error, $|e|_{avg}$, measured in each trial

$|e|_{avg}$(N)	Line	Ramp	Sin	Avg
Tactile	8.7	2.3	3.2	4.7
Visual	4.4	4.7	9.1	6.0
Tactile + Visual	5.4	4.4	7.0	5.6

Table 4.4.4 Cumulative absolute force error, $I_{|e|}$, measured in each trial

$I_{|e|}$(Ns)	Line	Ramp	Sin	Avg
Tactile	68.9	24.3	33.9	42.4
Visual	41.8	46.4	86.7	58.3
Tactile + Visual	60.7	43.4	118.9	74.3

4.4.7 Discussion

Our results showed that the user was able to complete all tasks within a reasonable amount of time and with $|e|_{avg}$ kept below the chosen force error tolerance. Thus, our pilot study has demonstrated the usability of our system. In addition, our data revealed some noteworthy observations. While one might expect that providing both tactile and visual force feedback should allow the user to perform best, results showed the contrary. Providing only tactile feedback yielded smaller force errors when compared to providing visual feedback or both tactile and visual feedback. A few possible explanations for these observations are discussed below.

Force is a haptic sense. Thus, feeding it back to the user's haptic sensory input is a closer mapping than converting it to a form for the user's visual sensory input. As a result, a haptic feedback signal may be more easily processed and understood by the user's sensory and cognitive systems. Furthermore, the pathways through which haptic and visual inputs are processed by humans and converted into a resulting reaction can be different.

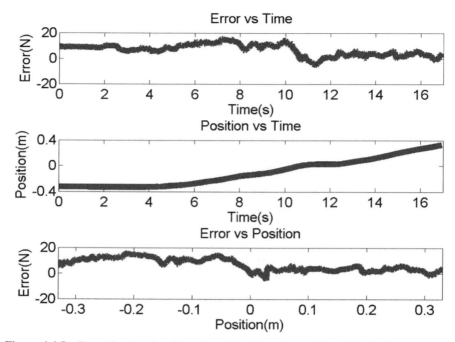

Figure 4.4.5   Example pilot data from the trial with tactile and visual feedback executing a sinusoidal force profile.

A visual sensory signal needs to first travel to the brain and be processed by the brain before it is converted to a motor neuron signal and sent to the target muscle to generate a reaction. A haptic signal may travel through a different route known as the reflex arc, in which the input signal to the sensory neurons travels to the spinal cord, gets processed at the spinal cord, where a motor neuron signal is then sent out to generate the reaction, without the input signal needs to reach the brain for processing first [37]. The reflex arc is shorter than the normal pathway through the brain, and thus, it has shorter reaction times. In our study, force feedback provided as visual arrows need to be processed by the user's brain first before motor signals are sent from the brain to the arm muscles to generate a corrective reaction for the applied force. A tactile signal may only need to travel through the reflex arc. Thus, these could be explanations of why haptic feedback produced better performance than visual feedback. A potential reason why providing visual feedback in addition to haptic feedback did not improve the performance could be that the added visual feedback is imposing an additional load on the user's visual sensory input and demanding more cognitive processing from the user. The system already provides through AR many other visual information related to trajectory creation and robot preview. Thus, adding visual force feedback may be overloading the visual channel.

4.4.8 Conclusion and Future Work

We have created a system utilizing augmented reality, gesture control, and haptic feedback for intuitive and effective robotic trajectory programming and execution. To our knowledge, this is the first in-situ untethered, handsfree system providing online control and feedback for both trajectory and force. We demonstrated the usability of our system showing that the user was able to program and execute different force profile trajectories. Preliminary results showed benefits for using haptic force feedback over visual force feedback or the combination of both.

In our next steps, we will be conducting full user studies to collect more data for validating our preliminary results. In addition, we will be testing the different components in a systematic way, including comparing different methods for trajectory teaching, different forms of robotic input, and overall performance of the system on executing different tasks. The outcome of our work will guide the design of human-robot interfaces utilizing the new technology of augmented reality towards achieving effective human-robot collaboration.

References

[1] F. Wallhoff, J. Blume, A. Bannat, W. Rosel, C. Lenz, and A. Knoll, "A skill-based approach towards hybrid assembly," *Advanced Engineering Informatics*, vol. 24, no. 3, pp. 329–339, 2010.

[2] C. A. Moore, M. A. Peshkin, and J. E. Colgate, "Cobot Implementation fo Virtual Paths and 3-D Virtual Surfaces," vol. 19, no. 2, pp. 347–351, 2003.

[3] R. Wilcox, S. Nikolaidis, and J. Shah, "Optimization of Temporal Dynamics for Adaptive Human-Robot Interaction in Assembly Manufacturing," in *Robotics: Science and Systems*, 2012.

[4] K. R. Guerin, C. Lea, C. Paxton, and G. D. Hager, "A framework for end-user instruction of a robot assistant for manufacturing," in *Robotics and Automation (ICRA), 2015 IEEE International Conference on*. IEEE, 2015, pp. 6167–6174.

[5] R. T. Azuma, "A survey of augmented reality," *Presence: Teleoperators and Virtual Environments*, vol. 6, no. 4, pp. 355–385, 1997.

[6] C. Breazeal, A. Edsinger, P. Fitzpatrick, and B. Scassellati, "Active vision for sociable robots," *IEEE Transactions on Systems, Man, and Cybernetics-Part A: Systems and Humans*, vol. 31, no. 5, pp. 443–453, 2001.

[7] B. C. Kress and W. J. Cummings, "11-1: Invited paper: Towards the ultimate mixed reality experience: Hololens display architecture choices," in *SID Symposium Digest of Technical Papers*, vol. 48, no. 1. Wiley Online Library, 2017, pp. 127–131.

[8] "Augmented reality and mixed reality," 2018, https://epson.com/moverio-augmented-reality.

[9] "Magic leap," 2018, https://www.magicleap.com/.

[10] X. Wang, S. Ong, and A. Nee, "A comprehensive survey of augmented reality assembly research," *Advances in Manufacturing*, vol. 4, no. 1, pp. 1–22, 2016.

[11] T. Engelke, J. Keil, P. Rojtberg, F. Wientapper, M. Schmitt, and U. Bockholt, "Content first: a concept for industrial augmented reality maintenance applications using mobile devices," in *Proceedings of the 6th ACM Multimedia Systems Conference*. ACM, 2015, pp. 105–111.

[12] S. Henderson and S. Feiner, "Exploring the benefits of augmented reality documentation for maintenance and repair," *IEEE Transactions on Visualization and Computer Graphics*, vol. 17, no. 10, pp. 1355–1368, 2011.

[13] S. Webel, U. Bockholt, T. Engelke, N. Gavish, M. Olbrich, and C. Preusche, "An augmented reality training platform for assembly and maintenance skills," *Robotics and Autonomous Systems*, vol. 61, no. 4, pp. 398–403, 2013.

[14] R. Sigrist, G. Rauter, R. Riener, and P. Wolf, "Augmented visual, auditory, haptic, and multimodal feedback in motor learning: a review," *Psychonomic Bulletin & Review*, vol. 20, no. 1, pp. 21–53, 2013.

[15] P. Larsson, D. Vastfjall, and M. Kleiner, "Ecological acoustics and the multi-modal perception of rooms: real and unreal experiences of auditory-visual virtual environments." *Georgia Institute of Technology*, 2001.

[16] A. Moon, D. Troniak, and B. Gleeson, "Meet me where I'm gazing: how shared attention gaze affects human-robot handover timing," in *International Conference on Human-Robot Interaction*, 2014, pp. 334–341.

[17] C. Perez Quintero, S. Li, C. Shing, W. Chan, S. Sheikholeslami, M. Van der Loos, and E. Croft, "Robot programming through augmented trajectories," in *VAM-HRI Workshop at The International Conference on Human-Robot Interaction*, 2018.

[18] S. A. Green, M. Billinghurst, X. Chen, and J. G. Chase, "Human-robot collaboration: A literature review and augmented reality approach in design," *International Journal of Advanced Robotic Systems*, vol. 5, no. 1, p. 1, 2008.

[19] Z. Pan, J. Polden, N. Larkin, S. Van Duin, and J. Norrish, "Recent progress on programming methods for industrial robots," *Robotics and Computer-Integrated Manufacturing*, vol. 28, no. 2, pp. 87–94, 2012.

[20] M. F. Zaeh and W. Vogl, "Interactive laser-projection for programming industrial robots," in *Mixed and Augmented Reality, 2006. ISMAR 2006. IEEE/ACM International Symposium on*. IEEE, 2006, pp. 125–128.

[21] J. W. S. Chong, S. Ong, A. Y. Nee, and K. Youcef-Youmi, "Robot programming using augmented reality: An interactive method for planning collision-free paths," *Robotics and Computer-Integrated Manufacturing*, vol. 25, no. 3, pp. 689–701, 2009.

[22] S. A. Green, J. G. Chase, X. Chen, and M. Billinghurst, "Evaluating the augmented reality human-robot collaboration system," *International Journal of Intelligent Systems Technologies and Applications*, vol. 8, no. 1–4, pp. 130–143, 2009.

[23] P. Kormushev, S. Calinon, and D. G. Caldwell, "Imitation learning of positional and force skills demonstrated via kinesthetic teaching and haptic input," *Advanced Robotics*, vol. 25, no. 5, pp. 581–603, 2011.

[24] F. Steinmetz, A. Montebelli, and V. Kyrki, "Simultaneous kinesthetic teaching of positional and force requirements for sequential in-contact tasks," *IEEE-RAS International Conference on Humanoid Robots*, vol. 2015-December, pp. 202–209, 2015.

[25] M. Minamoto, K. Kawashima, and T. Kanno, "Effect of force feedback on a bulldozer-type robot," in *2016 IEEE International Conference on Mechatronics and Automation, IEEE ICMA 2016*, 2016, pp. 2203–2208.

[26] A. Talasaz, A. L. Trejos, and R. V. Patel, "The Role of Direct and Visual Force Feedback in Suturing Using a 7-DOF Dual-Arm Teleoperated System," *IEEE Transactions on Haptics*, vol. 10, no. 2, pp. 276–287, 2017.

[27] B. Pier, P. Valentini, and M. E. Biancolini, "Interactive Sculpting Using Augmented- Reality, Mesh Morphing, and Force Feedback," no. March, pp. 83–90, 2018.

[28] B. M. Davis, "Effects of Visual, Auditory, and Tactile Navigation Cues on Navigation Performance, Situation Awareness, and Mental Workload, Tech. Rep. February 2007.

[29] J. B. van Erp, H. A. van Veen, C. Jansen, and T. Dobbins, "Way-point navigation with a vibrotactile waist belt," *ACM Trans. Applied Perception*, vol. 2, no. 2, pp. 106–117, apr 2005. [Online]. Available: http://dl.acm.org/citation.cfm?id=1060581.1060585

[30] D. Tam, K. E. MacLean, J. McGrenere, and K. J. Kuchenbecker, "The design and field observation of a haptic notification system for timing awareness during oral presentations," in *Proceedings of the SIGCHI Conference on Human Factors in Computing Systems – CHI '13*. New York, New York, USA: ACM Press, 2013, p. 1689. [Online]. Available: http://dl.acm.org/citation.cfm?doid=2470654.2466223

[31] M. K. X. J. Pan, J. McGrenere, E. A. Croft, and K. E. Maclean, "Exploring the Role of Haptic Feedback in an Implicit HCI-Based Bookmarking Application," *IEEE Transactions on Haptics*, vol. Submitted, pp. 1–12, 2012.

[32] Y. Yu, X. Wang, Z. Zhong, and Y. Zhang, "Ros-based uav control using hand gesture recognition," in *2017 29th Chinese Control And Decision Conference (CCDC)*, May 2017, pp. 6795–6799.

[33] A. Haddadi, E. A. Croft, B. T. Gleeson, K. MacLean, and J. Alcazar, "Analysis of task-based gestures in human-robot interaction," in *2013 IEEE International Conference on Robotics and Automation*, May 2013, pp. 2146–2152.

[34] "Thalmic labs," 2018, https://www.thalmic.com/.

[35] T. H. Michael, "Scientific computing: an introductory survey," *The McGraw-540 Hill Companies Inc.: New York, NY, USA*, 2002.

[36] C. P. Quintero, M. Dehghan, O. Ramirez, M. H. Ang, and M. Jagersand, "Flexible virtual fixture interface for path specification in telemanipulation," in *Robotics and Automation (ICRA), 2017 IEEE International Conference on*. IEEE, 2017, pp. 5363–5368.

[37] D. Kenneth S. Saladin, *Anatomy & Physiology: The Unity of Form and Function*. McGraw-Hill Education, 2014. [Online]. Available: https://books.google.ca/books?id=lmr8nQEACAAJ

4.5 Augmented Reality Instructions for Shared Control Hand-held Robotic System

Joshua Elsdon and Yiannis Demiris

Personal Robotics Laboratory, Imperial College, London, UK

This work includes two contributions. Firstly, the accuracy of an untrained human user completing a trajectory with a hand-held robot is characterized. We found that designers should expect up to 63 mm error in robot position and 0.18 rad (10°) error in orientation when the user is given augmented reality guidance. Secondly, we have demonstrated that providing augmented reality guidance can significantly improve the accuracy of speed regulation, position error and orientation error by 19.3%, 15.5%, and 39.2% respectively. This is compared to a situation when the user has full knowledge of the trajectory expected of them, but no substantial visualizations to guide them.

4.5.1 Introduction

Hand-held robotic systems offer a number of advantages over traditional stationery and mobile robots. Much of the bulk, complexity, and cost of articulation and mobility can be absorbed by the human user. This shift of complex tasks away from the robot can allow for the design of the robot to be more effective at the elements of the task where the robot adds the most value. For example, in the situation where the task is to apply a liquid coating to a large complex object. Rather than having a large gantry-based robot, or a complex and expensive mobile robot, a human can be used to deploy a hand-held robot to the correct location, and the precise task of application and measurement can be performed by the robot.

Such a system may be able to fill a niche between highly precise robotic arm based spraying robots, typically used on production lines, and highly trained technicians knowledgeable enough to complete one-off task effectively with manual spray equipment. A hand-held system would offer some of the precision of a robotic system and some of the flexibility of a manual spraying system.

However, in order to have a hand-held system work effectively the robot and the user must have a shared understanding of the task at hand. Head-mounted augmented reality is a promising technology for communicating information to a user whilst allowing them to work with as little hindrance as possible. This is because by default augmented reality systems allow the user to see the real world, allowing them to see potential hazards in the work environment, unlike a virtual reality system that must always take into account such hazards and make them available to the user in the visualizations shown to them. Additionally, for tasks demanding dexterity and hand-eye coordination, such as the spraying task we are considering, augmented reality allows those functions of the user to not be impaired. Allowing the designer to focus on improving the experience, rather than finding ways to avoid hindering it.

This work seeks to provide designers of augmented reality systems for hand-held robots an estimate of what degree of accuracy they can expect from the user, as well as providing some preliminary demonstration that augmented reality visualizations can improve the accuracy and quality of movements over unguided movements. This was achieved by conducting a preliminary study of 8 people conducting trajectories that were both guided and unguided. The system details can be seen in Figure 4.5.1.

4.5.1.1 Project background

This work is a continuation of our work on assisted spraying technologies. Previously we have presented work [1] on how to automate a single axis hand-held robot using a receding horizon approach. That work put the decision-making process in the hands of the human user, then the algorithm presented chose the best actuation strategy to apply the most liquid over a given time horizon. This approach can be effective, though it has a tendency to be greedy, making it difficult for the user to plan a global strategy. This was demonstrated in our recent work [2], where we performed a user study to measure the effectiveness of the approach. This study provided mixed results, some users found the assisted system cumbersome and found themselves fighting with the system. This work aims to provide the groundwork for shifting the decision-making capability from the user to the robot, with the expectation that this will cause less disagreement between the user and the robot. Further, agreeing on a plan of collaboration as presented in this work, could allow the development of algorithms for the hand-held spraying robot that takes a global approach to planning, rather than the greedy, receding horizon approach.

Figure 4.5.1 This figure shows a 3rd person view of a user performing a trajectory using the detailed version of the visualization. The hand-held robot and the mannequin are motion tracked by an infra-red camera system. This image was taken with the Microsoft HoloLens and would not normally be visible to a 3rd party. The HoloLens has been added into the image and foreground elements highlighted.

4.5.1.2 Related work

Gregg-Smith et al. [3] presented a range of experiments comparing the efficacy of various user feedback methods in aiding the user to position a hand-held robot at a given location. They tested a monocular augmented reality headset, a virtual reality headset, and a robot-mounted display, as well as a novel robot gesturing system. They concluded that all visual feedback methods performed similarly, both in regard to task completion and user task loading. However, this study also included a robot that could fully solve the task once in range of the target. This meant that the user was not required to perform any precise movements whilst using the robot. They did also present a non-robotic baseline, where the user must align a wand with the target position to an accuracy of 5 mm and 5° for 200 ms, though this stationary goal is not informative for our problem of tracking set trajectories.

4.5.2 Experimental Setup

The experiment consists of 8 trajectories that the users were asked to complete with the hand-held robot, both with a detailed visualization and with a basic one. The visualizations are shown in Figure 4.5.2. The aim of providing a basic visualization is to get some understanding in which ways a detailed visualization could detract from the quality of motion, though for the purposes of making comparisons the users need some confirmation of the path they are expected to follow. Hence the simple visualization is a minimalistic description of the direction they should follow. On the other hand, the detailed visualization shows the user the plane they are supposed to sweep with the gantry of the hand-held robot, where they should press and release the trigger and the speed at which they should be traveling. For all of the experiments the plane that they should sweep is the same distance from the mannequin. The mannequin is both physically present and rendered in the augmented reality system to provide confirmation that the calibration is working as expected. The speed was also the same for all trails, set at 30 cm/s. This consistency was designed such that the user can have a full

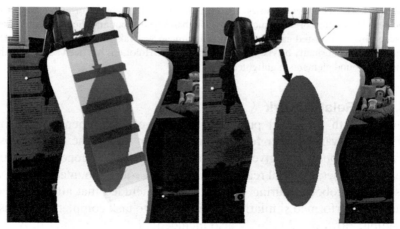

Figure 4.5.2 This figure shows both the detailed and simple visualizations used in the experiments. The detailed visualization has bars which move along the graphic at the speed that the user is expected to emulate. The user should aim to sweep the gantry of the hand-held spraying robot across the green section of the visualization, whilst maintaining orthogonality of the robot to the trajectory. The user is expected to do all of the same things with simple visualization. The simple visualization is only to help the user remember the direction they are expected to move the robot along. The blue area has no bearing on this work, though is defined in previous work [1, 2].

understanding of what trajectory they are expected to complete, even when there is only the simple visualization indicating the direction of the path to undertake.

Each of the participants were asked to complete 8 trajectories, where they completed each twice, alternating between the detailed and simple visualizations. Half of the participants undertook each trajectory with the detailed visualization first, the other half with the simple one. This interleaving of the two types of trial is to help ensure the user has a very good understanding of the parameters of the trajectories (speed, height above the mannequin, etc.) even when they are not shown these in the visualization.

The trajectory of the spraying robot is captured by an infrared camera-based motion capture system. The mannequin is tracked also so that the experimental area can be moved conveniently, though the mannequin was not moved during the trial for each user.

To ensure the visualizations are located accurately, virtual markers were manually placed on each of the motion tracking cameras. The reported location of the virtual markers was compared to the calibrated location of the cameras provided by the motion tracking software in a manner described by Ho et al. [4]. This method provides a least-squares approximation of the transform between the augmented reality coordinates and the motion tracking coordinates. Each of the virtual markers used the spatial anchor system provided by the Microsoft HoloLens. This ensured that they would track any changes in the coordinate system of the AR headset. Though not measured formally for this work, the accuracy of this calibration method is roughly 1–2 cm.

4.5.3 Results

The participants in this study were all familiar with the hand-held system and augmented reality headset used in this experiment and consisted of 2 females and 6 males. There are two categories of results that are of interest: relative quality measures and absolute quality measures. Relative quality measures do not reference the set trajectory, and absolute ones do. This distinction is important because it would be unreasonable to expect users to match parameters of a trajectory without being shown them, as is the case with the simple visualization. However, we can still analyze whether the movement that they did match the general criteria that were asked of them, namely, the path should be straight, at a constant speed and the robot should be orthogonal to the direction of travel at all times. These general criteria match

Table 4.5.1   A summary of the metrics analyzed, their standard deviations, and the p-value when considering the proposition "the error is lower in the case of the detailed visualization"

Error Metric	Units	Detailed	SD	Simple	SD	p-value
Abs. Position	mm	63.7	28.8	72.0	30.8	0.122
Rel. Position	mm	6.42	308	6.43	3.72	0.992
Trajectory Speed	mm/s	73.2	38.3	88.8	39.2	0.0255
Instant Speed	mm/s	157	51.6	178	62.4	0.0418
Abs. Orientation	rad	0.183	0.111	0.238	0.156	0.0265
Rel. Orientation	rad	0.314	0.129	0.350	0.146	0.136

the assumptions that our algorithm uses to find paths in our previous work [1]. All of the results here are summarized in Table 4.5.1.

Position Accuracy: The position accuracy was measured by taking the measured position of the robot and measuring the perpendicular distance to the trajectory. This is calculation is shown in Equation 4.5.1, where D is the distance from the line, S is the position vector of the start of the line, E the end, and P is the point under consideration. The absolute value of the distance was taken and averaged over the trajectory to arrive at the mean error from the trajectory. Participants performed better with the detailed visualization with an average error of 0.0636 m compared to 0.0720 m (p $=$ 0.12).

However, if we look at the error from the best-fit straight line of the user's trajectory, we see no difference between the visualization types, both diverting from the best fit line by an average of 6.4 mm. This shows that the visualization is not helping to keep the users traveling on a straight trajectory, trough it does help them stay in the vicinity of the target trajectory.

$$D = \frac{\left\| \vec{SE} \times \vec{PS} \right\|}{\left\| \vec{SE} \right\|} \tag{4.5.1}$$

4.5.3.1 Speed regulation

For all trials, the users were required to move the robot at 0.3 m/s. There are two metrics that are informative here, error in average speed over trajectory and speed error during trajectory.

The average speed of the robot ($S_{\text{trajectory}}$) was significantly more accurate with the detailed visualization, 0.073 m/s error, compared to the simple visualization, 0.089 m/s error (p $=$ 0.025). During the movement it was possible to see some variation in the speed as users were trying to match the moving bars in the visualization. If we look at the average error during the trajectory (S_{instant}), the detailed visualization performs better with 0.16 m/s error compared to 0.18 m/s (p $= 0.042$). The fact that the instantaneous speed error is significantly larger shows that the users are better at estimating the speed over the whole trajectory rather than keeping the correct speed at any given moment. The method of calculating the average speed error ($S_{\text{trajectory}}$) and instantaneous speed error (S_{instant}) is shown in Equations 4.5.2 and 4.5.3 respectively.

$$S_{\text{trajectory}} = \frac{1}{N} \left(\sum_{i=0}^{N} s_i \right) - s_{\text{target}} \qquad (4.5.2)$$

$$S_{\text{instant}} = \frac{1}{N} \sum_{i=0}^{N} \| s_i - s_{\text{target}} \| \qquad (4.5.3)$$

4.5.3.2 Orientation accuracy

The users were asked to keep the robot orthogonal to the direction of movement at all times, and in both versions of the visualization, the direction required is shown. Therefore, we can have two metrics to measure the performance of the users' alignment accuracy, the relative orientation of the robot in regard to its movement direction, and the alignment with the requested orientation. In both of these metrics the detailed visualization outperforms the simple visualization, 0.31 rad vs. 0.35 rad (p $= 0.13$) for the relative alignment, and 0.18 rad vs. 0.23 rad (p $= 0.026$) for the absolute alignment.

4.5.4 Conclusion

It can be seen that the more detailed visualization allowed the users to perform better in all of the metrics. Though this is not a particularly surprising result, the users had access to more information from the more detailed visualization. However, demonstrating the performance of the chosen visualization over that of a lesser visualization was not the aim of this preliminary work. Here we have demonstrated a baseline for user movements with the robot with effectively no guidance and demonstrated that even a somewhat simple visualization displaying key information helps the user rather than hindering them. It is hoped that a designer of a similar system can use the data provided here to allow them to design assistive algorithms that are using assumptions about the user's ability to comply with the instructions given. For example, an active head on such a spraying robot should be able to account for roughly 63 mm of deviation from the planned path and an orientation error of 0.18 rad (10 degrees), when the user has detailed information provided via an AR headset. However, this would increase to 72 mm position error and orientation error of 0.238 rad (13.6 degrees) error if provided with less convenient spatial cues.

Most of the remaining error in moving the robot through trajectories is likely difficulty perceiving depth and obstruction of the real world by the visualizations. Future work could attempt to provide visualized feedback to the user regarding their performance, helping to emphasis the mistakes that they are making. Further, assistive algorithms for hand-held robots can be improved with realistic knowledge of the capability of the human user to comply with trajectory requests.

References

[1] Joshua Elsdon and Yiannis Demiris. "Assisted painting of 3D structures using shared control with a hand-held robot". In: *2017 IEEE International Conference on Robotics and Automation (ICRA)*. 2017, pp. 4891–4897.

[2] Joshua Elsdon and Yiannis Demiris. "Augmented Reality for Feedback in a Shared Control Spraying Task". In: *2018 International Conference on Robotics and Automation (ICRA)*, in press.

[3] Austin Gregg-Smith and Walterio W. Mayol-Cuevas. "Investigating spatial guidance for a cooperative handheld robot". In: *2016 IEEE International Conference on Robotics and Automation (ICRA)*. 2016, pp. 3367–3374.

[4] Nghia Ho. *Finding Optimal Rotation and Translation Between Corresponding 3D points*. 2013. URL: http: //nghiaho.com/?page_id=671.

4.6 Augmented Musical Reality via Smart Connected Pianos

**Daniel M. Lofaro[1], Frank Lee[2], Edgar Endress[3]
and Chung Hyuk Park[4]**

[1]Department of Electrical and Computer Engineering,
George Mason University, USA
[2]College in the Media Arts and Design, Drexel University, USA
[3]School of Art, George Mason University, USA
[4]Department of Biomedical Engineering, School of Engineering and
Applied Science, George Washington University, USA

The chapter describes the proposed implementation of an augmented musical reality system that will help increase comradery through musical interaction, with a focus on comradery between people who are separated geographically and who hold different political, social, and financial backgrounds. Specific topics covered in this work are: (1) how to measure comradery and other emotions through musical interaction, (2) methods for audience interaction through reactive audio and visual feedback mechanisms, and (3) methods for keeping the real-time control fast enough, and latency low enough, to allow for musical interaction.

4.6.1 Introduction

Today's political and economic climate has increased the division between the "haves" and "have nots" as well as between people with differing political/social ways of thinking. These divide typically happen between people of differing social, racial, political, and/or economic backgrounds. All of the latter people are typically separated geographically and thus do not interact with each other on a day to day basis. These separations are typically by city and/or city section (i.e. center city, west city, suburbs, etc.). The goal of this project is to get these people interacting with each other using a common social mechanism, in this case music. This will be done by introducing smart connected pianos to urban and suburban environments creating an augmented musical reality that allows residence to play music with each other even if they are in different communities. Our hypothesis is that this interaction will help increase the connected communities' comradery and empathy for each other.

Based on the neurological and behavioral science that music is an effective pathway to impact on mental states [1–4] as well as physical behaviors [5–7], we plan on developing an inter-city augmented musical reality interaction framework that can affect people's way of thinking despite ones social, racial, political, and/or economic background. The combination has both technological and social dimensions and will help us answer our overarching question of "can remote physical interaction increase empathy?"

To answer this question, we will place six connected pianos throughout three distinct cities, Washington, D.C., Philadelphia, PA, and Fairfax, VA. These pianos will be outfitted with persistent real-time audio, visual, and haptic feedback via remotely actuated keys. Each smart connected piano will be connection to multiple other pianos allowing the user to play with people from other parts of the city and even other cities. In short, when a person presses a key on a piano in Washington D.C. the same key will move and play in Philadelphia and/or Fairfax.

This will encourage interaction with people with different backgrounds and create comradery between people that would not normally interact. To allow everyone to enjoy the work, the piano's music will be (1) streamed live on the internet (2) will have an interactive phone "app" where people from around the world can participate in playing the pianos, and (3) will have live interactive visual projections on structures around each piano where the interaction will affect the music and thus help include the audience in the collaborative remote music playing. The proposed public and university locations for the pianos are ● 30th street station and the ExCITe Center at Drexel University in Philadelphia, PA ● Union Station and the Autism & Neurodevelopmental Disorders Institute (ANDI) at George Washington University in Washington D.C. ● as well as Old Town Square and the Mason Innovation Exchange (MIX) at George Mason University in Fairfax, VA. An example of the Smart Connected Piano setup can be seen in Figure 4.6.1.

We also hypothesize that when humans work with other humans remotely, they will perform better at the given task, in this case, play music, and they feel more connected/empathy with the remote party. Additionally, we hypothesize that the correctness of the audio/visual feedback will improve the seamlessness of interaction between users at each piano location. Finally, we

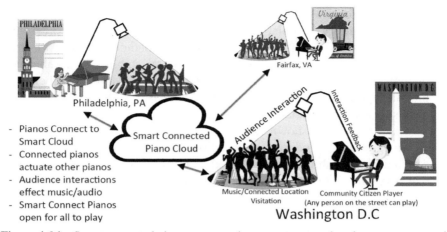

Figure 4.6.1 Smart connected piano concept. pianos are connected to the smart connected piano cloud and share key-movements, audio, and visualizations.

also hypothesized that the "better music" two parties play together will be an indicator of more comradery they feel towards each other.

The metrics and methods to determine the latter performance measures are stated in II-A. The data used to make the latter determinations will come from each of the connected pianos and their audience interface during the public deployment. Data collected will include • raw audio from the piano • key press note, velocity, and time from the piano's keyboard and • motion/input from the crowd interface. The latter data will be recorded and stored for later analyses on a remote system.

Our participating community stakeholders will help us with community engagement, event advertising, and smart connected piano location determination. Our first participating community stakeholder is The Arts Council of Fairfax, they will focus on the Washington D.C. Metro Area (D.C./Fairfax). Our second participating community stakeholder is the University City District (UCD), they will focus on the Philadelphia, PA area.

4.6.2 Methodology

4.6.2.1 Measuring emotion

Psychologically, one crucial aspect of social interactions is "empathizing" [8]. Empathizing is the capacity to attribute mental states, such as feelings, thoughts, and intentions to other people, and to respond to their mental states with an appropriate emotion [9–11]. A form of auditory stimulus, called rhythmic auditory stimulation (RAS), is well-established in neurological rehabilitation and therapy [12–14]. RAS is a method in which the rhythm functions as a sensory cue to induce temporal stability and enhancement of movement patterns by providing a temporal constraint for the patient's internal optimized path of motion. In this sense, RAS can be an effective means for inducing social engagement and emotional activities. Neurological studies [15–17] have shown that activity in the premotor cortex may represent the integration of auditory information with temporally organized motor action during rhythmic cueing. Based on this theory, researchers have shown that rhythmic auditory stimulation can produce significant improvements in physical activities [14, 18, 19]. However, we believe a higher-level form of stimulus, such as music, is required to stimulate the premotor cortex where strong relevance is found between neural domains for music, emotion, and motor behaviors. Given that music has shown such a long history of therapeutic effects on psychological [20–23] and physical problems [24, 25], we aim to measure the emotional expressions of a player (with musical emotional measures), remote player (with musical emotional analysis on the remote site), and the audience (with behavioral measures) to study the dynamics of social impact of music.

Musical emotion measures: Emotions are expressed through music via many different musical features, and many studies focused on Music Emotion Recognition (MER) have shown good progress. We aim to improve the real-time aspects to bring impacts on seamless human-system musical interactions. Our research focuses on improving prediction time by using fast and efficient feature extraction and machine learning methods to accurately extract features from live audio samples and classify the levels of arousal and valence in the signal to determine the emotion being conveyed. Both openSMILE [26] and Sinusoidal Transform Coding (STC) [27, 28] are being utilized to extract features from the 1000 Song Database and train our machine learning algorithms with annotated emotion data from the dataset.

Behavioral measures: To evaluate the participants' physical activities and social interaction, we will incorporate metrics from physical therapy and rehabilitation. For assessing the participants' gestures and small motions, we have determined from the literature that the best approach for our problem is to use the following metrics: range of motion (ROM), path length (PATH), peak angular velocity (PAV), movement time (MT), spatio-temporal variability (STV), and movement units (MUs). ROM is defined as the difference between the maximum and minimum angles during a trajectory and its increase is linked to an increase in the functional use of an afflicted arm [23, 29, 30]. PATH is defined as the 3-dimensional length of the path traveled by the hand and is said to reflect straightness of a reaching trajectory [30]. PAV is the maximum angular velocity that occurs during a trajectory. This is used as an indirect measure of force, of which an increase would be indicative more confident motion [30]. MT is the time required to move in one trajectory. STV refers to the variability of motion as it relates to time of which is comprised of temporal variability and spatial variability when correlated [23]. MUs defined in [30–32] are the number of peaks in the trajectory curvature. These metrics will allow us to characterize reaching movement to determine whether a treatment is effective or not.

Both the behavioral and musical emotion characteristics will be measured in two domains of activation (A) and valence (V) in Russell's circumplex model [33] as shown in Figure 4.6.2. We will incorporate Matlab and the OpenSmile [26] in the implementation of STC feature extraction and analysis models for audio (music) signals for emotion estimation and utilize vision sensors with OpenPose [34–36] to collect robust behavioral parameters and analyze the proposed parameters.

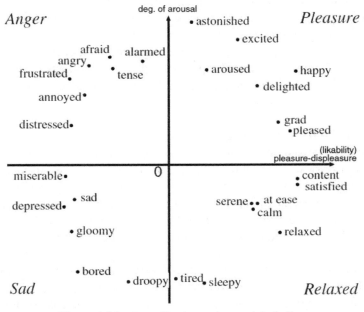

Figure 4.6.2 Russell's circumplex model of affect.

4.6.2.2 Visual feedback

The responsive visualizations at each city site will be projected onto a surface near the piano, in most cases the sidewalk or ground. At the first layer of response, the visualizations will react to the music being played in real-time, for example flashing or bouncing with each note. At the second layer of interaction, a camera will capture the projection space and input visual data into a live Unity environment that will interpret that data in real-time and then respond by changing the visualization. The effect of this arrangement will be that participating onlookers can manipulate and change the projection through their gestures. For example, a wave of the hand might push away a field of color, or a stomp of the foot will scatter a group of Koi fish swimming in a pond. Finally, a third layer of interaction is achieved through real-time emotional analysis of both the musicians and the crowd using Dr. Park's research on gestures and human affect, and those variables affect the visualizations on a global scale, for example, tinting the hue of the entire visualization red to represent anger or intensity of affect.

Visualizations will be built and coded in Unity, a game design software and engine that is quickly becoming industry standard for interactive development. The program will be a black box that responds to user inputs by outputting visual and audio elements into a display, in the same way any game responds to button or touchscreen inputs by outputting visual data onto a screen. In our case, the output display is a street projection, and the primary input data is a human gesture collected by a camera mapped to the projected display. Additional inputs include the piano keys, as well as affect and emotional data gathered based on the body language of the piano player and the various onlookers.

4.6.2.3 Real-time control

To achieve persistent real-time low-latency communication we will borrow from our prior work in cloud-robotics. In previous work we used a geographically adjacent cloud server to reduce the distance that the data has to travel, thus reducing latency [37]. The path that the data takes cannot be specifically defined, therefore we then implemented a bounty hunting methodology where we would request the same problem to be solved by multiple servers and only take the first server to response (i.e. the fastest). This resulted in a 99%+ reliability of real-time deadline data delivery when using five or more servers with between 85% and 90% on-time arrival probability [38].

For the augmented musical reality via smart connected pianos we need to get the data from point A to point B at 99%+ real-time low-latency delivery rate or the interaction will be negatively affected. The maximum amount of latency that musical participants can tolerate in a collaborative music task is varied. An experiment by Waldo showed that 60 percent of participants reached their latency limit at the 40 ms to 60 ms range, while the other 40 percent went well over the 80 ms range [39]. These findings coincide with the research of Barbosa and Cordeiro [40], as well as Boley and Lester [41], as they also established the 40 ms range to be the threshold for an optimal performance. This also explains Kon and Lago's [42] findings that a propagation latency of 30 ms will be unnoticeable to the audience since this latency amount is still tolerable. The latter gives us our limitations for round trip latency for our smart connected piano system.

We will reduce latency by (1) reducing the distance the data needs to travel by "promoting" a desired shortest distance routing path by connecting through multiple geographically adjacent servers that approximately form a line between point A (Smart Piano 1) and point B (Smart Piano 2) (similar to our geographically adjacent server approach); and (2) using multiple short paths to increase the probability that one will arrive on time (similar to our bounty-hunting server approach). This will result in a network of geographically adjacently connected servers see Figure 4.6.3.

Each smart connected piano will connect to its geographically adjacent servers. The bounty hunting method will be used to transmit the data. This means that the data will be sent over multiple paths and only the data that is received first will be used and the rest will be ignored. We expect to see an in 99%+ reliability of real-time deadline data delivery when using five or more signal paths when each path has an on-time delivery probability of

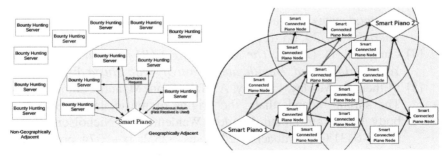

Figure 4.6.3   (LEFT) Geographically adjacent server setup. Smart piano is connected to servers closest to it. (RIGHT) Geographically adjacently connected bounty hunting server system shows how each of the geographically adjacent servers are connected to each other and share information/data with each other. Geographically shortest path shown in RED.

between 85% and 90%. The latter is based on the results of our prior work [38]. We also expect to find that our increase in real-time reliability is at the cost of using more bandwidth.

4.6.3 Conclusion

The work above described the proposed implementation of an augmented musical reality system that will help increase comradery through musical interaction, with a focus on comradery between people who are separated geographically and who hold different political, social, and financial backgrounds. As stated, this will be done by introducing smart connected pianos to urban and suburban environments allowing residence to play music with each other even if they are in different communities. The specific methodologies proposed in this work were (1) how to measure comradery and other emotions through musical interaction, (2) methods for audience interaction through reactive audio and visual feedback mechanisms, and (3) methods for keeping the real-time control fast enough, and latency low enough, to allow for musical interaction. As stated, this is the plan for a large scale augmented musical reality system that will be implemented in the coming years.

References

[1] Storr, A. (2015). *Music and the Mind*. Simon and Schuster.
[2] Wan, C. Y., Rüüber, T., Hohmann, A. and Schlaug, G. (2010). The therapeutic effects of singing in neurological disorders, *Music Perception: An Interdisciplinary Journal*, vol. 27, no. 4, pp. 287–295.
[3] Magee, W. L. and Davidson, J. W. (2002). The effect of music therapy on mood states in neurological patients: a pilot study. *Journal of Music Therapy*, vol. 39, no. 1, pp. 20–29.
[4] Campbell, D. (2000). *The Mozart Effect [R] for Children: Awakening Your Child's Mind, Health, and Creativity with Music*. ERIC.
[5] Mithen, S., Morley, I., Wray, A., Tallerman, M. and Gamble, C. (2006). The singing neanderthals: the origins of music, language, mind and body, by steven mithen. London: Weidenfeld & Nicholson, 2005. isbn 0-297-64317-7 hardback£ 20 & us $25.2; ix+ 374 pp. *Cambridge Archaeological Journal*, vol. 16, no. 1, pp. 97–112.
[6] Ziv, N., Granot, A., Hai, S., Dassa, A. and Haimov, I. (2007). The effect of background stimulative music on behavior in Alzheimer's patients. *Journal of Music Therapy*, vol. 44, no. 4, pp. 329–343.

[7] Rauscher, F., Shaw, G., Levine, L., Wright, E., Dennis, W. and New-comb, R. (1997). Music training causes long-term enhancement of preschool childrens spatial–temporal reasoning. *Neurological Research*, vol. 19, no. 1, pp. 2–8.

[8] Decety, J. and Jackson, P. L. (2006). A social-neuroscience perspective on empathy. *Current Directions in Psychological Science*, vol. 15, no. 2, pp. 54–58.

[9] Baron-Cohen, S. and Wheelwright, S. (2004). The empathy quotient: an investigation of adults with asperger syndrome or high functioning autism, and normal sex differences. *Journal of Autism and Developmental Disorders*, vol. 34, no. 2, pp. 163–175.

[10] Eisenberg, N. (1989). Empathy and sympathy.

[11] Harris, J. C. (2003). Social neuroscience, empathy, brain integration, and neurodevelopmental disorders. *Physiology & Behavior*, vol. 79, no. 3, pp. 525–531.

[12] Kwak, E. E. (2007). Effect of rhythmic auditory stimulation on gait performance in children with spastic cerebral palsy. *Journal of Music Therapy*, vol. 44, no. 3, pp. 198–216.

[13] Malcolm, M. P., Massie, C. and Thaut, M. (2009). Rhythmic auditorymotor entrainment improves hemiparetic arm kinematics during reaching movements: a pilot study. *Topics in Stroke Rehabilitation*, vol. 16, no. 1, pp. 69–79.

[14] Thaut, M., Kenyon, G., Schauer, M. and McIntosh, G. (1999). The connection between rhythmicity and brain function. *IEEE Engineering in Medicine and Biology Magazine*, vol. 18, no. 2, pp. 101–108.

[15] Kohler, E., Keysers, C., Umilta, M. A., Fogassi, L., Gallese, V. and Rizzolatti, G. (2002). Hearing sounds, understanding actions: action representation in mirror neurons. *Science*, vol. 297, no. 5582, pp. 846–848.

[16] Molnar-Szakacs, I. and Overy, K. (2006). Music and mirror neurons: from motion to emotion. *Social Cognitive and Affective Neuroscience*, vol. 1, no. 3, pp. 235–241.

[17] Rizzolatti, G. and Craighero, L. (2004). The mirror-neuron system. *Annual Reviews in Neuroscience*, vol. 27, pp. 169–192.

[18] del Olmo, M. F. and Cudeiro, J. (2005). Temporal variability of gait in parkinson disease: Effectsof a rehabilitation programme based on rhythmic sound cues. *Parkinsonism & Related Disorders*, vol. 11, no. 1, pp. 25–33.

[19] Pacchetti, C., Mancini, F., Aglieri, R., Fundaro, C., Martignoni, E. and Nappi, G. (2000). Active music therapy in parkinson's disease: an

integrative method for motor and emotional rehabilitation. *Psychosomatic Medicine*, vol. 62, no. 3, pp. 386–393.

[20] Madsen, C. K. (1997). Emotional response to music. *Psychomusicology: A Journal of Research in Music Cognition*, vol. 16, no. 1–2, p. 59.

[21] Nayak, S., Wheeler, B. L., Shiflett, S. C. and Agostinelli, S. (2000). Effect of music therapy on mood and social interaction among individuals with acute traumatic brain injury and stroke. *Rehabilitation Psychology*, vol. 45, no. 3, p. 274.

[22] Castillo-Perez, S., Gomez-Perez, V., Velasco, M. C., Perez-Campos, E. and Mayoral, M. A. (2010). Effects of music therapy on depression compared with psychotherapy. *The Arts in Psychotherapy*, vol. 37, no. 5, pp. 387–390.

[23] Siedliecki, S. L. and Good, M. (2006). Effect of music on power, pain, depression and disability. *Journal of Advanced Nursing*, vol. 54, no. 5, pp. 553–562.

[24] Karageorghis, C., Jones, L. and Stuart, D. (2008). Psychological effects of music tempi during exercise. *International Journal of Sports Medicine*, vol. 29, no. 7, pp. 613–619.

[25] Jing, L. and Xudong, W. (2008). Evaluation on the effects of relaxing music on the recovery from aerobic exercise-induced fatigue. *Journal of Sports Medicine and Physical Fitness*, vol. 48, no. 1, p. 102, 2008.

[26] Eyben, F., Wollmer, M. and Schuller, B. (2010). Opensmile: the munich versatile and fast open-source audio feature extractor. in *Proceedings of the 18th ACM International Conference on Multimedia*. ACM, 2010, pp. 1459–1462.

[27] Quatieri, T. F. and McAulay, R. J. (1992). Shape invariant time-scale and pitch modification of speech. *IEEE Transactions on Signal Processing*, vol. 40, no. 3, pp. 497–510.

[28] Kim, J. C., Rao, H. and Clements, M. A. (2014). Speech intelligibility estimation using multi-resolution spectral features for speakers undergoing cancer treatment," *The Journal of the Acoustical Society of America*, vol. 136, no. 4, pp. EL315–EL321.

[29] Brooks, D. and Howard, A. M. (2010). A computational method for physical rehabilitation assessment," in *Biomedical Robotics and Biomechatronics (BioRob), 2010 3rd IEEE RAS and EMBS International Conference on*. IEEE, 2010, pp. 442–447.

[30] Chen, Y. P., Kang, L. J., Chuang, T. Y., Doong, J. L., Lee, S. J., Tsai, M. W., Jeng, S. F. and Sung, W. H. (2007). Use of virtual reality

to improve upper-extremity control in children with cerebral palsy: a single-subject design. *Physical Therapy*, vol. 87, no. 11, pp. 1441–1457.

[31] Fetters, L. and Todd, J. (1987). Quantitative assessment of infant reaching movements. *Journal of Motor Behavior*, vol. 19, no. 2, pp. 147–166.

[32] Hobson, R., Ouston, J. and Lee, A. (1989). Naming emotion in faces and voices: Abilities and disabilities in autism and mental retardation. *British Journal of Developmental Psychology*, vol. 7, no. 3, pp. 237–250.

[33] Russell, J. A. (1980). A circumplex model of affect. *Journal of Personality and Social Psychology*, vol. 39, no. 6, p. 1161.

[34] Cao, Z., Simon, T., Wei, S. E. and Sheikh, Y. (2017). Realtime multiperson 2d pose estimation using part affinity fields. in *CVPR*.

[35] Simon, T., Joo, H., Matthews, I. and Sheikh, Y. (2017). Hand keypoint detection in single images using multiview bootstrapping. in *CVPR*.

[36] Wei, S. E., Ramakrishna, V., Kanade, T. and Sheikh, Y. (2016). Convolutional pose machines, in *CVPR*, 2016.

[37] Lofaro, D., Asokan, A. and Roderik, E. (2015). "Feasibility of cloud enabled humanoid robots: Development of low latency geographically adjacent real-time cloud control," in *Humanoid Robots (Humanoids), 2015 15th IEEE-RAS International Conference on*.

[38] Lofaro, D. M. and Asokan, A. (2016). Low latency bounty hunting and geographically adjacent server configuration for real-time cloud control, in *2016 IEEE International Conference on Robotics and Automation (ICRA)*, pp. 5277–5282.

[39] Greeff, W. (2017). The influence of perception latency on the quality of musical performance during a simulated delay scenario, Ph.D. dissertation, University of Pretoria.

[40] Barbosa, A. and Cordeiro, J. (2011). The influence of perceptual attack times in networked music performance, in *Audio Engineering Society Conference: 44th International Conference: Audio Networking*. Audio Engineering Society.

[41] Boley, J. and Lester, M. (2007). The effects of latency on live sound monitoring, in *Audio Engineering Society, Convention Paper*, p. 20.

[42] Lago, N. and Kon, F. (2004). The quest for low latency. in *ICMC*.

5

Enactive Steering of Simulations for Scientific Computing

Brandon Mechtley*, Julian Stein, Todd Ingalls, Sha Xin Wei and Christopher Roberts

School of Arts, Media, and Engineering, Arizona State University, USA
E-mail: bmechtley@asu.edu
*Corresponding Author

5.1 Introduction

This chapter will discuss several case studies in developing composition software and media systems for *enactive steering* of computational models. Drawing on past works in computational steering, experiential media systems, and responsive environments, we make use of an enactivist framework to guide the development of simulations that can be steered in real-time, that is in parallel with execution, using whole-body movement by both individuals and ensembles of unbounded size, providing new opportunities for scientific communication, hypothesis formation, and decision-making.

In particular, we will discuss a media system known as *EMA*, or *An Experiential Model of the Atmosphere*, as well as the media composition framework used in its development, *SC*. SC, in addition to being a framework for composition of responsive media environments, can also be used as a suite of scientific computing tools to provide real-time mappings between different computational sensing and immersive feedback modalities and a live simulation, serving as easily-understood real-time analogs to common data visualization paradigms. We have developed SC and systems such as *EMA* as potential drop-in components of a scientific computing pipeline, where existing models can be connected to them, ideally making the only work for investigators that of choosing the appropriate mappings.

EMA is part of a larger research stream on creating computational platforms for integrated, gestural interaction with complex models via multi-modal interfaces that will allow for fluid human-in-the-loop control of simulated scenarios. The main challenge to developing such a system is handling new densities of data that approach a continuous distribution. Our strategy is that an effectively continuous dynamical systems approach can provide principles for designing a system able to evolve in real-time to non-discrete, multi-user gestural control of rich experiential scenarios that tap embodied, human experience [1–6]. Thus, we seek to develop computational paradigms that will allow designers to leverage the full potential of the increasing density of sensors and computational media in everyday situations by providing a wholly experiential means of controlling and interacting with dense sensors and media. One way to scaffold our technical and design imaginary is to use the model of the swimming pool in place of the model of a graph. How does the water coordinate its activity with the activity of its inhabitants and the wind blowing across its surface? Many forces modulate its movement and condition. Some forces are due to people swimming through the water, pushed by and pushing the liquid surrounding them. Others are due to the waves on its surface, or currents distinguished by momentum, or in the case of the deep ocean, salinity, and temperature. Still, others are due to the wind which acts continuously across a continuous surface – the continuously extended interface between the air and the water. Note that whereas we may regard a rock thrown into the water or a swimmer as a compact, point-like source of motive force, aggregates of entities or even more essentially, extended continuous fields do not fit this model of an atomic agent. Dyadic (1–1) relational interaction is a small, sparse subset of much richer fields of experiential dynamics. Thus, we seek a more ample way to conceive engagement between different fields of media in a responsive environment.

To fit with this paradigm of responsive media, our method of choosing appropriate simulations has been to look for computational adaptations of continuous (e.g. differential geometric or topological) models to the scientific analysis of dense, heterogeneous environments like weather systems and urban spaces. These continuous models complement discrete models (e.g., discrete graphs) of procedural computation processes. We adopt techniques from signal processing and computer science that are also shared with machine perception, fault-tolerant systems, or autonomous systems but we do so with the distinctive intent to keep human-in-the-loop control of the experience that can give designers computational paradigms leveraging *collective, embodied* experience [1, 2, 7–10]. The three classes of continuous models we investigate are (1) homogeneous generalized computational physics of materials, (2) continuous evolution of metaphorical states, and (3) heterogeneous atmospheric models, such as models that mix for example agent-based

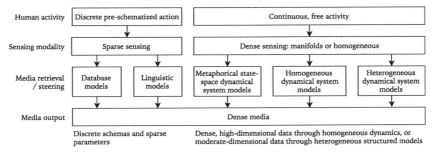

Figure 5.1 Layered activity tracking and computational media processing for experiential environments [38].

models of urban dynamics, models of geophysics, or rule-based systems that model interventions by large scale sociopolitical institutions, that condition the lattice models of atmosphere itself. Figure 5.1 elaborates on this model of dense media, demonstrating how discrete and continuous sensing modalities can contribute to significantly different experiences in media systems.

We first lay the framework for embodied, enactivist [1, 2, 7, 11] approaches to the design of computer interfaces, and more generally of responsive environments augmented by computation. In that context, we will define what we mean by real-time, multimodal, whole-body, gestural and multi-person engagement with an immersive responsive environment.

5.1.1 Experiential Media Systems

An important motivation and context for our work is the focus on the whole experience, in the senses of William James [12], contemporary phenomeno-logical work on experience [10, 13–17] and movement-based experience [18–20]. Under these approaches, experience cannot be decomposed into a finite number of perceptual or functional component dimensions and reassembled in some linear superposition of independent features. Senses of rhythm and of mathematical pattern are examples of such *apperceptions*. Despite this irreducibility of experience, this non-decomposability of experi-ence into "independent" sensory dimensions, there are useful means of ascer-taining accounts of experience that can be shared objectively across instances: notably methodological and experimental approaches by Petitmengin [21], Sha [22, 23], and Bregman [24].

Francisco Varela, Evan Thompson, and others introduced the notion of *enactive* experience to describe how we progressively construct our sense of, concepts, and know-how about the material world through engagement and empirical experience: "We propose the term enactive to emphasize the growing conviction that cognition is not the representation of a given world by a pregiven mind but is rather the enactment of a world and a mind on the basis of a history of the variety of actions that a being in the world performs" [1]. We extend that cognitivist sense of *enaction* to a more thoroughly processualist one of how subjects, organisms [25], technical ensembles [26], more generally any individuals and their environment co-construct each other [4, 27, 28] via structural interaction [29, 30].

Experimental platforms scaffolding such whole experiences – *experiential systems* and *responsive environments* – have been built by Sundaram [31, 32], Sha [33], and others (see survey on responsive environments in [34]). By *embodiment* we mean sense-making which is conditioned on one's corporeal engagement with the material world. By *material,* we mean the union of physical, energetic, social, and affective fields [4, 35].

Experiential systems, then, are immersive, multimodal, real time, and multi-person. Immersivity can be more precisely framed in the phenomenological distinction between acting, being, sensing in the world without any reflection – *thrownness* (*geworfenheit*), versus the state of being reflexively aware of one's stance with respect to the world (called "defamiliarization" or *Verfremdungseffekt* in some technical contexts [36]). In this more precise sense, being immersed in a situation is independent of the sensory modalities that are being most exercised. One can be immersed in reading a book on one hand, and on the other, be largely "clinically" disengaged even in a full-body, physical interaction.

Our experiential systems are designed multimodally, that is, the software framework for our experiential systems, *SC*, is designed for integrated, gestural interaction with complex models via multi-modal interfaces that allow fluid human-in-the-loop control of densities of data that approach a continuous distribution. The system evolves in real-time to non-discrete, multi-user gestural control of rich experiential scenarios, which tap embodied, human experience [1–6, 37]. Thus, we leverage the full potential of the increasing density of sensors and computational media in everyday situations by providing a wholly experiential means of controlling and interacting with dense models.

Finally, our responsive environments are all designed for multi-person use, which requires a different sort of design than extrapolating from the design of "single-user interaction" where a single user is seated in front of a screen with keyboard and mouse WIMP interfaces that can only be controlled by one person at a time. This is a concrete setting for designing for human-human and human-system interaction based on ensemble experience and on ensemble activity. Concretely, ensemble interaction concerns situations where there are three or more human participants so that we do not fall back on social conventions encoded in dyadic interaction. Also, this sidesteps human-machine interaction design that is implicitly predicated on single-user WIMP interface design including web document interfaces and most non-game "applications," whether on mobile or desktop computers. A simple example of ensemble engagement is a group walking in a circle to stir up the atmosphere or the ocean model to form a large vortex, as in Figure 5.2.

5.1.2 Steerable Scientific Simulations and Abductive Method

Creative experimental scientific work relies on constructing fresh instruments of observation *in tandem with* fresh theoretical interpretations of freshly observed phenomena. We call this on-the-fly co-construction of theory, instrumentation and observation, which is characteristic of creative work in science as well as other disciplines, the *abductive* method [39–41].

Some computational science applications have also adopted human-in-the- loop modulation of parameters through the use of computational steering. In computational steering of simulations, investigators change parameters of computational models on the fly and immediately (or as close to immediately as possible) receive feedback on the effect, in parallel with the execution of the simulation. In practice, computational steering allows investigators to quickly explore alternative paths of evolution of system state, such as by introducing exogenous changes to boundary conditions or simulation parameters.

Computational steering has been applied to the real-time control of scientific simulations, such as fluid dynamics [42] in general, air safety [43], flood management [44, 45], particle physics [46], astrophysics [47], and cardiology [48], and several frameworks have been created for integrating the methodology into new and existing simulations deployed on high-performance computing platforms, including SCIRun [49], RealityGrid [50], and WorkWays [51].

We conceptually extend the notion of computational steering to real-time human-in-the-loop modulation of any computationally modulated environment where the results are immediately perceived, thus minimizing the time between configuration and analysis of a simulation. With advances in dense sensing modalities and experiential media, previous responsive media

(a) A trio forms warm clouds in a simulation mapping optical flow to an increase in water vapor and temperature. On the scrim, the condensed liquid water mixing ratio is sonified by segmenting the field into eight vertical strips and mapping the field averages and differences to amplitudes and filter properties of a sample-based audio instrument.

(b) A group walks in a ring around a pivot to simulate a hurricane.

Figure 5.2 Ensembles ($n \geq 3$) steering a realtime simulation by coordinated whole-body interaction. B. Mechtley, M. Patzem, and C. Rawls. Synthesis 2018.

systems have expanded upon these primarily screen-based computational steering interactions in several aspects:

(1) Embodied, enactive environments allow comparatively unconstrained engagement with the computation, such as through full-body movement or the use of physical props or other aspects of the environment. For example, gestural input can afford more degrees of freedom to

allow multiple parameters to be controlled simultaneously and physical props can be used to construct detailed geometry for more varied, non-parametric boundary conditions.

(2) Applications range widely from basic experiential experiments (e.g. relation between memory and corporeal movement, or rhythmic entrainment of ensembles of people and time-based media processes) to artistic installations and performance (e.g. Serra [52] and Timelenses [53]).

(3) Designing for whole body and ensemble engagement implies *thick* [5], multimodal, analog and digital engagement with the environment as opposed to interacting along one or a few dimensions of sensory perception.

In studying the use of responsive media environments for computational scientific investigation, our key interest lies in observing what types of behaviors in these systems could contribute to scientific practice, such as rapidly "sketching" hypotheses, perhaps in advance of more numerically reproducible studies. With lowering costs and advances in computing resources, both in HPC systems and desktop hardware, many of the simulations that now require HPC systems may eventually be able to be steered at responsive rates, so we study the use of those models which can currently be simulated at these rates to get an understanding of where responsive interaction can fit into future computational science workflows.

5.2 EMA: An Experiential Model of the Atmosphere

As an initial exploration, EMA is installed in the Intelligent Stage ("iStage") space in Synthesis at the School of Arts, Media, and Engineering at Arizona State University: a 30×30-foot black box space with a sprung dance floor, theater grid with 16 DMX-controllable RGB LED theatrical lights and additional floor-mounted lights, 4 K floor and vertical scrim laser projections, horizontal ceiling-mounted and vertical infrared-filtered cameras, infrared light emitters, floor-mounted contact, and boundary microphones, grid-mounted microphones, and 8.2-channel surround audio. The space is designed to be modular to support multiple responsive environments with flexible HD video routing and Dante-compatible audio routing hardware.

As a responsive, steerable model of warm cloud physics, EMA satisfies many of the objectives of dense computational media that can be steered through collaborative human gesture and physical configuration of the space. Additionally, using a physical model of atmospheric dynamics allows us to explore human interaction with a simulated physical model that leverages participants' existing physical intuition of matter, exhibits phase changes, and can simulate phenomena at different spatial and temporal scales, all contributing to a rich set of processes and forms that can be studied by investigators in the space.

EMA implements an incompressible fluid flow model along with additional computation of buoyancy, condensation, and evaporation of water vapor, and thermodynamics. During each timestep of the simulation, the model allows for external video textures to manipulate the simulated fields, including air velocity, pressure, water vapor, liquid water, temperature, and viscosity. Global scalar parameters can also be manipulated in real-time, such as ground pressure and temperature, altitude, spatial and temporal scale, specific heat capacities of dry air and water vapor, external wind speed and direction, and gravity magnitude and direction. For mathematical and implementation details, see [38].

Note that in EMA, it is possible to condition experiments using physical, "dumb" props: that is, physical materials that are not embedded with any particular "smart" tracking hardware. Since computer vision is used not to reduce a scene into a collection of separate objects, but rather to segment it based on material properties (for example, amount of IR absorption or visual patterns), people can construct props of any shape and size using everyday materials. See Figure 5.3 for some examples. Figure 5.4 shows the detailed interface the investigator can use to parameterize the simulation, including physical constants, spatial physical properties such as gravity, external sources of wind, location of the ground plane, and mappings between sensed activity and simulated fields in the model, such as forcing wind velocity through optical flow or mapping presence of objects and bodies as sources of air pressure or water vapor.

5.2.1 Visualization

The simulation's fields can then be viewed with a number of different visualization modes, including conventional pseudocolor images given a colormap, particle flow fields, tracer particles, line integral convolution, vector feather plots, and additional artistic renderings composed of multiple fields, such as temperature and different phases of water. The base set of visual mappings has been developed to reflect conventional scientific visualizations from familiar platforms such as Matplotlib, ParaView, and MATLAB. Each of these visualization modes can be seamlessly interchanged using a mobile tablet interface, and EMA supports layering multiple visualizations, such

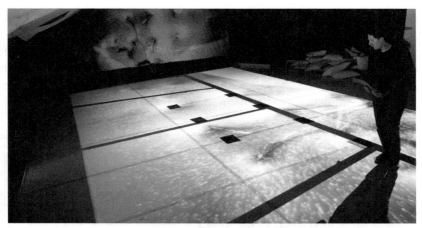

(a) A researcher studies flow between multiple low-pressure regions by placing sheets of paper, whose silhouette is mapped to a constant reduction in pressure. Particles flow continuously with the velocity field, depicted by a dense pseudocolor plot. Select tracer particles are sonified as spatialized voices.

(b) Two researchers study changes to a simulated vortex sheet by narrowing a slit between two poles. The angle of air velocity is mapped to hue, with a constant external source of wind coming from upstage. Velocity magnitude orthogonal to the wind is sonified by averaging the velocity field over a coarse grid and spatializing the sound according to the grid cell centers.

Figure 5.3 Realtime corporeal interaction with dense, high-dimensional, GPU-accelerated simulation of atmospheric dynamics. On the order of 100 sound processes provide spatialized sonic textures with a palpable landscape for enactive, embodied engagement.

as being able to view flow lines or specific tracer particles on top of a composite rendering of air temperature, water vapor, and condensed liquid water. The tablet interface also allows viewers to adjust scaling parameters, color maps, and compositing *in situ* without the need to return to a desktop interface in order to encourage all investigative activity to occur embedded within the simulation environment.

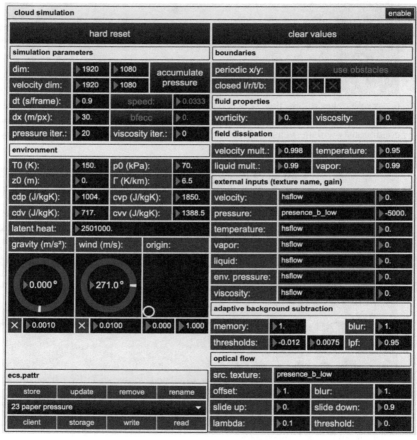

Figure 5.4 Top: Investigators can quickly change simulation variables, re-map named OpenGL textures to simulation fields, and manage presets from both a desktop and mobile tablet interface without needing to leave the simulation environment. **Bottom:** Simulation fields can be visualized using conventional techniques such as pseudocolor mapping and line integral convolution for two-dimensional vector fields.

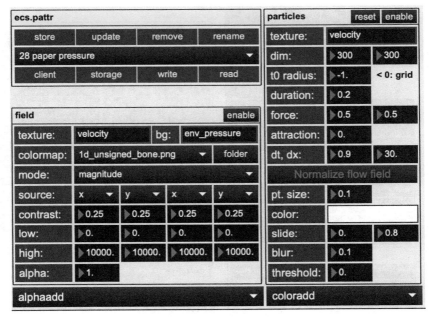

Figure 5.4 Continued.

5.2.2 Sonification

The sonic affordances of the space can also be used to communicate important dynamics within the simulation that may be difficult to attend to visually, such as spatialized activity of airflow or the position and velocity of moving tracer particles. We have implemented two modes of sonification in the environment that allow investigators to sonify activity within regions and particular points in space. In particular, a field-based sonification tool allows multiple participants to scale a bounding rectangle around their bodies or static objects to sonify the dynamics of specific regions of space, as in Figure 5.5. The underlying field is then subdivided into a variable number of zones, and the average changes of the field within the zones are then sent through a multi-channel sample player and filterbank and ambisonically spatialized around the participants [38].

In a separate particle-based sound synthesizer, individual tracer particles are simulated in the model, which follows the velocity of the air. Each particle is mapped to a separate voice, and its speed and direction are mapped onto

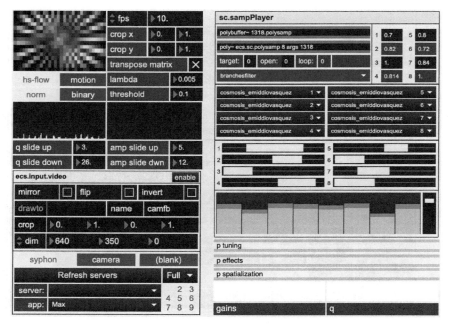

Figure 5.5   Input video routed from any computer, including all simulated fields, can be mapped to a tuneable bank of resonant bandpass filters. In this example, incoming video is split into eight segments, and their optical flow and intensity are mapped onto center frequencies and gains for eight banks of filters applied to loop samples, which are then diffused across the loudspeaker array. Additional properties allow for tuning the pitch and reverberation of each audio stream.

different aspects of the synthesized sound. To allow designers or investigators to choose informative sound textures, they are able to select an audio file or recorded audio sample which is then sampled with a granular synthesizer. The angle of velocity of each tracer particle is then mapped to the center frequency of a resonant bandpass filter, and the speed of the particle is mapped to the particle's volume. This mapping is particularly effective at sonifying sudden changes in particle velocity, such as when it enters a vortex or suddenly encounters a gust of wind. When a particle is in circular motion, for example, the synthesized voice will make repeating sweeps up and down in frequency content. Each particle is then ambisonically spatialized within the space using the *SPAT* ambisonic spatialization tool [54] to allow participants to understand whole-field dynamics when they are visually focused on a particular region of the simulation.

5.2.3 Enactive Scenarios

The basic dimensions of our enactive scenarios are the number of people and props (physical manipulates), the weather phenomenon being simulated and experienced, the tools or instruments for inspecting or modulating the state of the simulation.

As earlier mentioned, we distinguish between the experience of one, two, or ensemble ($n \geq 3$) of people co-constructing an experience in realtime with the steerable environment. Thus, the experiments on how the simulation is experienced are designed differently and accordingly. For each of three scenarios, we list recorded experimental behavior from open-ended sessions working with the model as a solo investigator, a pair, a pair using objects in the lab space to construct experiments, and as a guided ensemble. In the ensemble scenario, people dispensed with instruments and used their bodies to walk in a coordinated way to steer the simulation holistically. These scenarios include:

Cloud formation and airflow on a horizontal plane: a horizontal simulation with a ceiling-mounted camera is constructed where each square pixel corresponds to 900 m^2 of simulated space, the simulated ambient temperature is 150 Kelvin, and motion of entities in the space is mapped to an increase in water vapor, which condenses nearly instantaneously.

Cloud formation on a vertical plane: a simulation with a vertically oriented camera facing an opposing vertical projection surface is constructed where each pixel corresponds to 100 m^2 of simulated space and the lapse rate of ambient temperature with altitude is 6.5 K/km with a sea-level temperature of 288.15 K, resulting in a temperature gradient ranging from 288.15–241.35 K. Presence of bodies and objects in the space acts as an obstruction to fluid flow, while movement is mapped to an increase in water vapor and temperature, causing buoyant lift and eventual condensation, usually slightly above-head when participants are standing approximately 5 feet from the projection.

Cloud formation and airflow with wind on a vertical plane: a simulation parameterized similarly to the previous scenario, but an external, constant source of downstage velocity (wind) is added, allowing participants to observe the effects of airflow around themselves and objects.

Table 5.1 summarizes a sampling of participatory scenarios we have tested using the system. [55] contains full details of participant observation trials, where increasing the number of participants in the space was seen to increase joint expressive capabilities, such as through actions as coordinated movement and manipulation of instruments and sharing objects or physical space in the eye of the camera. Inclusion of physical instruments in the space, ranging from isolated objects such as pipes and rope to furniture, such as stools and tables, allows participants to construct stable fluid boundaries or affect larger and more complex regions of the simulation than they could with their bodies alone, such as through spinning objects overhead or jointly moving large objects between each other.

Table 5.1   Observed experimentation strategies in three simulations

Fluid Flow and Cloud Formation on a Horizontal Plane	Cloud Formation on a Vertical Place	Cloud Formation and Air Flow with Wind on a Vertical Plane
One person		
Spinning with arms outstretched to produce vortices, clouds.	Swaying side-to-side to introduce buoyant water vapor and heat and watch clouds form.	Raising one or both hands to obstruct wind above lifted condensation level.
Using sheets of paper, mapped to low pressure centers, to direct air flow.	Raising one or both hands and moving them parallel to the screen to introduce water vapor above the lifted condensation level.	
Walking in circles to produce vortices.	Walking parallel to the projection to leave a trail of buoyant water vapor.	
Two people		
Walking together in circles to produce more stable vortices	Holdinig or overlapping hands and moving hands down to leave a large trail of buoyant water vapor.	Standing in order of shortest to tallest or visversa to observe flow up or down an irregular slope.
Walking towards each other with arms outstretched to collide opposing fronts.		Using side-by-side hands at different heights to create irregular slops.
Two people with instruments		
Manipulating pipes to create an obstruction with a variable-size slit or channel to observe effects on vortex sheets.	Rotating and moving a vertically oriented foamtube horizontally to produce large areas of buoyant water vapor.	Rotating a large foam rectangle to observe flow up a straight incline.

Table 5.1 Continued

Fluid Flow and Cloud Formation on a Horizontal Plane	Cloud Formation on a Vertical Place	Cloud Formation and Air Flow with Wind on a Vertical Plane
Shaping an airfoil with a rope and observing effects on fluid pressure and velocity on either side.		Placing a sheet of dark paper close to the camera to observe flow up a straight incline.
Creation different curves and shapes with a rope at an oblique angle to simulated wind and observing fluid flow along their surface.		
Moving circular tables and stools into the space to observe flow around many objects.		
Spinning a wand overhead to create vortices.		
Ensemble $n \geq 3$		
Walking together in circle to produce a stable vortex.	Producing cloud masses as concurrent effect.	Producing cloud masses as concurrent effect.
Participating in group discussion and manipulation of several objects (sheets of blac paper, rope, metal tubes) to experiment and discuss results of different configurations on fluid flow.		

5.3 The SC Responsive Media Library

To facilitate the creation of responsive media environments, we have written a structured set of software abstractions, SC, that simplifies the physical sensing of environments and control of media instruments within them. It is intended to allow modular design that enables designers to rapidly pass signals between sensors, data transformations, and media; manipulation and production of dense, continuous media that produce a rich palette of possible media states; continuous evolution of media that can facilitate production of dynamic, responsive, rather than static or repetitive media environments; and designing at the level of metaphor that focuses the attention of developers on the intended effect of media states rather than underlying mathematical representations.

SC is designed to be used by event composers or installation artists who are not software engineers. The scripting language Max/MSP/Jitter, the lingua franca for live media scripting, permits the blending of decades of toolkits for live media processing from developers worldwide, such as cv.jit computer vision [56], MuBu multibuffer audio signal processing [57], SPAT spatialized audio [54], and Wekinator machine learning [58].

5.3.1 Architecture

The SC software frameworks are built around media instruments: code that transforms a media stream (typically sound, video, or light) into another media stream at real-time rates under the threshold of perceptible latency.

Media instruments are redefined continuously by parameters that are functions of features derived from sensor inputs, internal conditions, or the metaphorical state of the event. In a typical responsive environment, a suite of instruments process the time-based field media by mapping multimedia streams to each other.

Within the SC framework, nearly 300 utilities and instruments have been created to compose responsive environments, consisting of physical sensing and actuators, acoustic sensing and multi-channel audio, lighting arrays, and streaming video capture and visual projections. SC supports working between real-time media formats by using a standardized modular architecture whereby messages, including real-time streams of manipulated media as well as statistical calculations are passed between components. In this way, media can be easily interchanged or transmuted. To support this style of development, SC has been implemented in the real-time media programming environment Max/MSP/Jitter to make all layers of control legible in a common scripting language accessible to composers, designers, and scientists who do not need to be intimately familiar with the software engineering details and wish to work across all computational media types. In the following sections, we will describe a few example objects from each category: audio, video, lighting, physical, and intermediate simulations. In addition to the media instruments, SC provides utilities for manipulating data streams, including objects for scaling, easing, mapping, interpolating, and ramping signals as well tools for event detection, time-series analysis, and tools for producing sequences of data, such as Markov random processes and physical models.

5.3.2 Audio Instruments

SC's extensive collection of audio utilities includes several modules for audio effects and filtering, and reverberation effects to simulate different acoustic environments, granular synthesis, synthesis of parameter trajectories and sequences for pitches and frequency bands, a multi-band filterbank instrument that allows movement within a dense stream of video to manipulate audio streams, and several utilities for producing audio in different multi-channel setups aided by IRCAM's SPAT audio spatialization package [54].

5.3.3 Video Instruments

The collection of video utilities and instruments encompass everything from grabbing video from different sources to computer vision algorithms for feature extraction and post-processing of graphical output. SC has abstractions that wrap methods of reading data from cameras, files, and the Syphon [59] framework. A cornerstone of the SC package is the array of computer vision abstractions. These patches cover methods of background subtraction, edge detection, optical flow, and presence detection. All of these techniques have been built as GLSL fragment shaders to allow for higher resolution tracking while still increasing performance in comparison to traditional CPU-based versions. In addition to methods of quantifying video inputs, SC is also capable of visualizing the results from simulations. Two notable methods are through the use of particle systems and vector fields. The particles can either act as tracers, being guided by the output field of a simulation, or behave as agents, feeding data back into the system that is moving them. Vector fields reveal trends of change within a simulation's output by drawing poles pointing in the direction of the current delta.

SC also provides methods of modifying visualizations through post-processing effects. The list includes blooms, blurs, palette-based recoloring, and temporal shifts. A "time-space" effect produces an image consisting of delays on a per-pixel basis. The system stores incoming frames of video into a buffer and uses a mask to determine where in the buffer to sample for each individual pixel. This temporal processing method allows for a visual history of any video stream and also opens up opportunities for artistic use of delays.

5.3.4 Lighting Instruments

SCs lighting abstractions rationalize accessing and controlling DMX lights or other hardware via a common messaging protocol. Pre-built abstractions handle everything from scaling incoming data along the response curve of individual lights to communicating the correct DMX address with the chosen interface. This allows composers to blend in a state-of-the-art professional theatrical lighting to produce the highest quality visual qualia.

5.3.5 Physical Sensors and Actuators

The library of physical sensing patches supplies methods of reading and parsing data from different sensor configurations and controlling physical media. It includes modules for communicating with a network of WiFi-enabled development boards that control arrays of ultrasonic atomizers, interface with a custom bend-sensing glove interface, and parse data from x-io Technologies XOSC I/O board, which transmits IMU data and the state of up to ten additional sensors wirelessly via UDP bundled in the Open Sound Control protocol [60].

5.3.6 The SC State Engine: Continuously Evolving Media

To produce continuously evolving behaviors within responsive environments using the SC instruments, we have created a state engine utility that allows designers to determine how the environment will evolve in response to the sensed state of the environment and current state of the media instruments or underlying simulations. The purpose is to meaningfully evolve the behavior of the entire ensemble of rich media instruments suggested by a continuous model of state evolution. We use a continuous dynamical system in place of Boolean, procedural, or stochastic (probabilistic) logics that are commonly implemented using finite state machines to provide rich behavior that responds to arbitrary fine nuance, and evolves robustly to arbitrary, even unanticipated, activity. Figure 5.6 diagrams the difference between responsive media systems where mappings between sensing (e.g. video, audio, wearables) are mapped directly to media instruments and systems in which a continuously evolving state engine acts as an intermediary, where sensors inputs drive the system state, which in turn directs the behavior of the system.

This continuous state evolution model is not a finite state machine, because the formalism admits continuous ranges of change and unbounded continua of possible states: a state can be the formal combination of any number of ingredient states. The state engine does not describe the state of component code or physical devices, which we call parameterization or presets. Rather, in our terminology, a state refers to a metaphorical description of the event as experienced by the inhabitants in the environment during a live

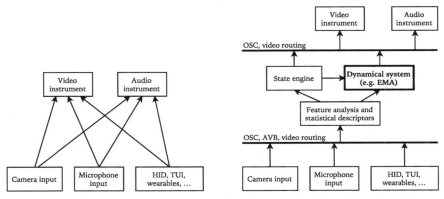

Figure 5.6 **Left**: An architecture for an audio-video experiential media system that uses manual, preprogrammed mappings between sensing modalities and audio and video instruments. **Right**: In comparison, SC allows for architectures using high-level descriptors for potentially multi-modal features to drive a state engine that guides transitions between states in audio and video instruments. As an example, EMA can both guide audiovisual instruments but also be guided by the state engine.

event. For example the state of an event could be characterized by terms such as the beginning, nighttime dormancy, people are bored, or stormy, which are nominally associated by the composer to combinations of features that can be derived from sensors (cameras, microphones, photocells, piezoelectric sensors, etc.). This interpolation of a state evolution layer in between the sensor data and the parameters controlling the software/ hardware media instruments allows the composer to design rich behavior while at the same time freeing her/him from locking that behavior design into a particular technology or particular technical instantiations.

It is also important to emphasize that the behavior is neither a fixed sequence (fixed tree of locally determined linear sequences of action) nor random (stochastic). More profoundly, the designer does not determine what actually happens, but rather the way the environment will tend to evolve in response to any activity. Thus, the state evolution acts on the space of potential, not actual activity. Nonetheless, the designer can condition the behavior as precisely and narrowly as desired. In practice, this system is best for medium to coarse qualitative changes in the behavior of arbitrarily rich complex environments, whereas specific action-response logic can be written using conventional ad-hoc code.

In practice, SC's state engine is implemented as an interface where designers can (a) define a number of states, (b) assign nominal sensor values to each state, and (c) bind parameters of the media instruments to these states. The designers can then arrange the states in a simplicial complex, allowing for both linear movement between fundamental states and more complex regions in the potential state space where the current state may be a combination of many fundamental states. (There is nothing sacred about which states are fundamental and which are (convex) sums of fundamental states). The systems state evolves as a function of both the incoming sensor features and an intrinsic dynamic based on minimizing an energy functional.

The vector distance of current sensor data from the nominal sensor vectors assigned to each fundamental state contributes to the energy of the current state of the environment. The physical model then evolves so as to minimize this energy by adjusting the position of the player state. In addition, several parameters allow designers to give each state a certain amount of static energy to fine-tune the contribution of the state to the movement of the player state. An example of state topology leading an environment through different times of day and seasons is shown in Figure 5.7.

In this topology, the state of the media environment can move between several metaphorical states associated with different seasons of the year and times of day. Within each season, the system state can evolve continuously between the different times of day, but significant seasonal state transitions will only occur overnight. In this image, the media environment is experiencing an overnight transition between summer and fall. If this topology were to drive EMA, for example, each state could be associated with different physical variables, such as average ground temperature or humidity, associated with different seasonal climates. The state engine is implemented in a graphical interface that makes it easy for designers to arrange the state topology, name states, and assign sensor data values and media parameters to each state.

5.4 Moving Beyond Point-and-Click Data Visualization

Responsive, enactive steering provides many opportunities for scientific communication, discovery, and decision-making. With the widespread availability of high-definition room-scale virtual reality platforms, data visualization will be a natural candidate for these now commonplace immersive media systems. However, many of these systems still make use of point-and-click metaphors from traditional graphical user interfaces. What immersive, real-time media systems afford beyond an extra dimension for visualization, however, are interactivity, multisensory feedback, and gestural control. Responsive media environments provide a rich repertoire of strategies for creating immersive

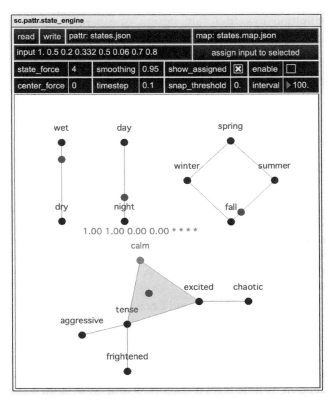

Figure 5.7 A dynamical state engine allows multiple system states to evolve continuously in response to activity within an environment. In this example, four state topologies of humidity, diurnal and seasonal cycles, and emotional states evolve concurrently to modulate different aspects of a responsive media environment.

environments that can reach beyond more commonly ocular centric notions of scientific communication. In addition, these environments have been built to allow for the type of ad hoc, improvisational activity – that is, play – that is necessary for creating the potential for discovery in computational models beyond pre-programmed scenarios or other fixed modes of interaction, such as through virtual cursors or other appropriations of two-dimensional WIMP metaphors. In the systems presented in this chapter, we have attempted to illustrate these possibilities. While An Experiential Model of the Atmosphere presents a highly idealized, simplified atmospheric model compared to global climate models or other HPC simulations, it has provided a suitable testbed for developing new means of steering, and we hope that it provides some inspiration toward future design of steerable simulations and data visualization systems.

References

[1] Varela, F. J., Thompson, E. and Rosch, E. (1991). The Embodied Mind: Cognitive Science and Human Experience. MIT Press, Cambridge, Mass.

[2] Dourish, P. (2001). Where the action is: the foundations of embodied interaction. MIT Press, Cambridge, Mass.

[3] Sha, X. W. (2009). Calligraphic video: Using the body's intuition of matter. International Journal of Creative Interfaces and Computer Graphics 1 (1, Inaugural Issue).

[4] Sha, X. W., Freed, A. and Navab, N. (2013). Sound design as human matter interaction. In: CHI '13 Extended Abstracts on Human Factors in Computing Systems. CHI EA '13, New York, NY, USA, ACM 2009–2018.

[5] Sha, X. W. (2013). Poiesis and Enchantment in Topological Matter. MIT Press, Cambridge, MA.

[6] Ingalls, T. (2013). Affect in media: Embodied media interaction in performance and public art. MultiMedia, IEEE 20(2): 4–7.

[7] Merleau-Ponty, M. and Landes, D. A. (2011). Phenomenology of perception. Routledge, Abingdon, Oxon; New York.

[8] Casey, E. S. (2007). In: Borders and Boundaries: Edging into the Environment. Volume Merleau-Ponty and Environmental Philosophy: Dwelling on the Landscapes of Thought, eds. S. L. Cataldi and W. S. Hamrick. Albany: SUNY Press.

[9] Casey, E. S. (2009). Getting back into place: toward a renewed understanding of the place-world. 2nd edn. Studies in Continental thought. Indiana University Press, Bloomington.

[10] Morris, D. (2004). The sense of space. SUNY series in contemporary continental philosophy. State University of New York Press, Albany (2004) 2004045291 David Morris. ill.; 24 cm. Includes bibliographical references (pp. 197–213) and index.

[11] Magnussen, T. (2014). Of epistemic tools: musical instruments as cognitive extensions. Organised Sound, 168–176.

[12] James, W. (1912). Essays on Radical Empiricism. Dover (Harvard 1912) (2003 (1912)).

[13] Gendlin, E. T. (1997). Experiencing and the creation of meaning: a philosophical and psychological approach to the subjective. Northwestern University Press, Evanston, Ill. Eugene T. Gendlin. 23 cm. Northwestern University studies in phenomenology and existential philosophy Originally published: Glencoe: Free Press of Glencoe, c1962. With new pref.

[14] Casey, E. S. (1997). The fate of place: a philosophical history. University of California Press, Berkeley.

[15] Casey, E. S. (1979). Perceiving and remembering. Review of Metaphysics 32: 407–436.

[16] Casey, E. S. (2000). Remembering: A Phenomenological Study. 2nd edn. Indiana University Press.

[17] Petitmengin, C. (2014). In: Researching the dynamics of intuitive experience. Volume Handbook of Research Methods on Intuition, ed. Marta Sinclair. Boston: Edward Elgar Publishing, 188–198.

[18] Sheets-Johnstone, M. (2012). From movement to dance. Phenomenology and the Cognitive Sciences 11(1): 39–57.

[19] Sheets-Johnstone, M. (1999). The Primacy of Movement. John Benjamins Publishing Company, Amsterdam.

[20] Sheets-Johnstone, M. (1981). Thinking in movement. Journal of Aesthetics and Art Criticism 39(4): 399–407.

[21] Petitmengin, C. (2014). In: Le corps du savant á la source du sens. Volume Epistémologie du corps du savant, ed. M. Quidu. Paris: L'Harmattan 188–198.

[22] Wei, S. X. and Gill, S. (2005). Gesture and response in field-based performance. In Proceedings of the 5th Conference on Creativity & Cognition 205–209.

[23] Sha, X. W. (2002). Resistance is fertile: Gesture and agency in the field of responsive media. Configurations 10(3): 439–472.

[24] Bregman, A. (2008). ASA at McGill, Keynote CIRMMT, CIRMMT, McGill University.

[25] Longo, G. and Montévil, M. (2014). Perspectives on Organisms: Biological time, Symmetries and Singularities. Springer-Verlag.

[26] Simondon, G. (2016). On the mode of existence of technical objects. Univocal. Minnesota (2016 (1958)) translated by Cecile Malaspina and John Rogove.

[27] Scott, D. (2014). Gilbert Simondon's Psychic and collective individuation: a critical introduction and guide. Edinburgh University Press.

[28] Combes, M. (2013). Gilbert Simondon and the philosophy of the transindividual. Technologies of lived abstraction. MIT Press, Cambridge, Mass. (2013) 2012013224 Simondon. English Muriel Combes; translated, with preface and afterword, by Thomas LaMarre. 24 cm. Includes bibliographical references (pp. 109–119).

[29] Maturana, H. R. (2002). Autopoiesis, structural coupling and cognition: A history of these and other notions in the biology of cognition. Cybernetics and Human Knowing 9(3–4) 5–34.

[30] Maturana, H. R. and Varela, F. J. (1980). Autopoiesis and cognition: the realization of the living. Boston studies in the philosophy of science; v. 42. D. Reidel Pub. Co., Dordrecht, Holland; Boston (1980) Humberto R. Maturana and Francisco J. Varela; with a pref. to "Autopoiesis" by Sir Stafford Beer. 23 cm. Includes index.

[31] Sundaram, H. (2013). Experiential media systems. ACM Transactions on Multimedia Computing, Communications, and Applications. 2(3).

[32] Rikakis, T., He, J., Sundaram, H., Qian, G. and Spanias, A. (2005). An interdisciplinary arts and engineering initiative for experiential multimedia. In: ASEE Annual Conference and Exposition, ASEE 8715–8724.

[33] Sha, X. W. (2008). Human-computer interaction series. In: Poetics of Performative Space. Springer, London 549–566.

[34] Bullivant, L. (2006). Responsive Environments: architecture, art and design. Paperback edn. Victoria and Albert Museum.

[35] Massumi, B. (2002). Parables for the Virtual: Movement, Affect, Sensation (Post-Contemporary Interventions). Duke University Press.

[36] Jameson, F. (2000). Brecht and Method. Verso, London.

[37] Baran, M., Lehrer, N., Siwiak, D., Chen, Y., Duff, M., Ingalls, T. and Rikakis, T. (2011). Design of a home-based adaptive mixed reality rehabilitation system for stroke survivors. In: Proceedings of the 33rd Annual International Conference of the IEEE Engineering in Medicine and Biology Society. 7602–7605.

[38] Mechtley, B., Stein, J., Roberts, C. and Wei, S. (2017). Rich state transitions in a media choreography framework using an idealized model of cloud dynamics. In: Thematic Workshops 2017 – Proceedings of the Thematic Workshops of ACM Multimedia 2017, co-located with MM 2017.

[39] Morris, D. (2005). Bergsonian intuition, husserlian variation, peirceian abduction: Toward a relation between method, sense, and nature. The Southern Journal of Philosophy 43: 267–298.

[40] Peirce, C. S. In: Pragmatism as the Logic of Abduction (Lecture VII of the 1903 Harvard lectures on pragmatism). Volume Essential Peirce v. 2. 226–241.

[41] Psillos, S. (2000). In: Abduction: Between Conceptual Richness and Computational Complexity. Volume Abduction and Induction: Essays on their Relation and Integration, A. K. Kakas and P. Flach (eds.). Dordrecht: Kluwer 59–74.

[42] García, M., Duque, J., Boulanger, P. and Figueroa, P. (2015). Computational steering of CFD simulations using a grid computing environment. International Journal on Interactive Design and Manufacturing (IJIDeM) 9(3):235–245.

[43] Slotnick, J. P., Khodadoust, A., Alonso, J. J., Darmofal, D. L., Gropp, W. D., Lurie, E. A., Mavriplis, D. J. and Venkatakrishnan, V. (2014). Enabling the environmentally clean air transportation of the future: a vision of computational fluid dynamics in 2030. Philosophical Transactions of the Royal Society A: Mathematical, Physical and Engineering Sciences 372(2022): 20130317–20130317.

[44] Konev, A., Waser, J., Sadransky, B., Cornel, D., Perdigão, R. A., Horváth, Z. and Groller, M. E. (2014). Run watchers: Automatic simulation-based decision support in flood management. IEEE Transactions on Visualization and Computer Graphics 20(12):1873–1882.

[45] Cornel, D., Konev, A., Sadransky, B., Horváth, Z., Gröller, E. and Waser, J. (2015). Visualization of object-centered vulnerability to possible flood hazards. Computer Graphics Forum 34(3): 331–340.

[46] Sheharyar, A. (2014). Parallelization of computational steering of the electron avalanches in the high-energy particle detector simulators. Masters thesis, Qatar University.

[47] Xiao, J., Zhang, J., Yuan, Y., Zhou, X., Ji, L. and Sun, J. (2015). Hierarchical visual analysis and steering framework for astrophysical simulations. Transactions of Tianjin University 21(6): 507–514.

[48] Zhang, S., Xie, S., Wang, K. and Yuan, Y. (2015). Multi-scale Cardiac Electrophysiological Simulation: A GPU-based System by Computational Steering. (Iset) 160–163.

[49] Parker, S. G. and Johnson, C. R. (1995). Scirun: A scientific programming environment for computational steering. In: Proceedings of the 1995 ACM/IEEE conference on Supercomputing, ACM 52.

[50] Brooke, J. M., Coveney, P. V., Harting, J., Jha, S., Pickles, S. M., Pinning, R. L. and Porter, A. R. (2003). Computational steering in

realitygrid. In: Proceedings of the UK e-Science All Hands Meeting. Volume 16.

[51] Nguyen, H. A., Abramson, D., Kipouros, T., Janke, A. and Galloway, G. (2015). Workways: interacting with scientific workflows. Concurrency and Computation: Practice and Experience 27(16): 4377–4397.

[52] Khintirian, O. S. (2016). Serra vegetal life (http://serracreation. weebly.com/).

[53] Stein, J., Khintirian, O. S., Ingalls, T. and Sha, X. W. (2016). Time lenses apparatus (http://rhythmanalysis.weebly.com/time-lenses-apparatus.html).

[54] Carpentier, T. (2018). Une nouvelle implémentation du Spatialisateur dans Max.

[55] Mechtley, B., Roberts, C., Stein, J., Nandin, B. and Wei, S. X. (2018). Enactive steering of an experiential model of the atmosphere. In: Proceedings of the International Conference on Virtual, Augmented, and Mixed Reality, Springer, Cham. 126–144.

[56] Pelletier, J. (2013). cv.jit Computer Vision for Jitter. http://jmpelletier. com/cvjit/ 15.

[57] Schnell, N., Robel, A., Schwarz, D., Peeters, G., Borghesi, R., et al. (2009). Mubu and friends – assembling tools for content-based real-time interactive audio processing in max/msp. In: ICMC.

[58] Fiebrink, R. and Cook, P. R. (2010). The wekinator: a system for real-time, interactive machine learning in music. In: Proceedings of The Eleventh International Society for Music Information Retrieval Conference (ISMIR 2010) (Utrecht).

[59] Butterworth, T. and Marini, A. (2019). Syphon (http://http://syphon. v002.info).

[60] Freed, A. and Schmeder, A. (2009). Features and future of open sound control version 1.1. In: New Interfaces for Musical Expression (NIME).

6

Improving User Experience in Virtual Reality

6.1 Presence and Performance in the TRUST Game

**Fotis Liarokapis[1], Victoria Uren[2], Panagiotis Petridis[2]
and Adekunle Adeniyi Ajibade[3]**

[1]Masaryk University, Czech Republic
[2]Aston University, UK
[3]Coventry University, UK

Serious games that employ virtual environments are increasingly used in education and training. Presence, the feeling of immersion in a virtual world, is considered to be an important element of such games. This paper reports a study that examined whether interface factors that contribute to presence also affect players' perceived performance in the TRUST game. It was found that frustration is correlated to performance, and that ability to manipulate the interface and ability to focus are in turn, correlated to frustration.

6.1.1 Introduction

Serious games represent the state-of-the-art in the convergence of electronic gaming technologies with instructional design principles and pedagogies.

By combining the cutting-edge technologies with the sophisticated pedagogical models and theories, serious gamers have tackled areas from corporate training and education through to emergency medical response [1]. Many serious games exploit virtual environments which are widely credited with enhancing the user experience. In this paper, we report an evaluation of a serious game, TRUST, developed as part of the Voice Your View project [2] looking at aspects of presence and performance in a game designed to encourage democratic involvement in the care of urban environments.

The broadest definition of a serious game is, perhaps, a game played for a purpose other than entertainment. Zyda [3] provides a broad-stroke definition of a serious game as "a mental contest, played with a computer in accordance with specific rules that uses entertainment to further government or corporate training, education, health, public policy, and strategic communication objectives". Serious Games are games designed with the purpose not just to entertain, but to also solve a problem. These games can enhance learning by creating engaging and highly immersive experiences for the participants [4, 5]. Other benefits of serious games are:

- Scalability of game environments to large global communities
- Adapt to user requirements
- Closer modeling of user behavior
- Behavioral change
- Flow, feedback, visual and actual realism leading to higher levels of immersion
- Increased motivation and engagement
- Multimodal integration of interfaces and other technologies (e.g. AI, haptics, biofeedback, sensors networks, etc.) from convergence, mash ups, and user/community interaction

The key attributes of such games involve rules of motivation, known as Self-Determination Theory (SDT). SDT focuses on three interrelated categories: mastery, relatedness, and autonomy, which address the need to allow innate growth and wellbeing tendencies to flourish [6, 7]. Serious games come in different forms such as digital games, virtual worlds, board games, simulations simple web-based solutions, online virtual environments, mixed reality games, etc.

6.1.2 Examples

The use of virtual worlds, as a component of serious games is well-established, with flight simulators being the most generally known. In order to provide context for the TRUST game, this section describes some examples of serious games that use virtual worlds to teach participants about environmental issues and behaviors.

6.1.2.1 Ectopia

Ecotopia [8] is a serious game designed for 12+ audience funded by the Conservation International. It is free to play, and it builds on the popular city-building simulation model popularized by games such as Sim City. The game encourages environmentally conscious behavior in-game and links with Facebook, in order to create a social interaction around environmental issues. An interesting crossover between game and real-life happens when gamers send in accounts of real-life green measures. They get rewarded with in-game points, which also boost their profile in the social community as green heroes.

6.1.2.2 Anno 2070

Anno 2070 [9] is a city-building and management simulation game developed by Ubisoft. The game was released in late 2012. Like previous games in Anno series, players will settle a city and manage series of small islands in a region. The game presents several elements addressing environmental issues and raising awareness. There are two factions in Anno 2070. Each faction represents its own technology and lifestyle. The first faction is called Tycoon. This faction can be compared to present human population in developed countries. Tycoon citizens enjoyed luxurious commodities and lifestyle at expense of the environment. Tycoon's technology is focused on rapid expansion. Most of their technologies involve rapid exhaustion of natural resources and have negative impacts on the environment. The second faction is called Ecos. This faction can be compared to the environmental conscious population. Citizens of this faction promote sustenance lifestyle and most of their foods consist of vegetarian diets (with an exception of milk products). Ecos citizens emphasized on clean technology such as the use of solar and wind energy to produce electricity.

6.1.2.3 iSeed

The Living Stories Project (iSeed game) [10] is a social, online, location-based, alternate reality game, utilizing user-generated content including photos and stories brought together through mashups, using Facebook, Twitter, Google Maps and Second Life. This game aims to combine the real world with the online community by using social sites (i.e. Facebook and Twitter) and the Virtual worlds by using Second Life. The concept behind the game is to generate an interactive community wherein players share information and learn from each other, in accordance with social learning principles. By sharing this information, we aim to highlight the areas of work of the Eden Project, allowing new revenue to be created, and additionally to increase the environmental awareness of the players.

6.1.2.4 The code of everand

Dunwell et al. [11] present an evaluation of a game-based approach seeking to improve the road safety behavior amongst children aged 9–15 within the UK, made available outside of a classroom context as an online, browser-based, free-to-play game. The Code of Everand game was made available from November 2009 to November 2011, accessible through a player embedded within a web page. As such, the interface was constructed to be operated using a mouse on a desktop PC. Prior to launch, it was promoted through online search and display advertising on child-oriented websites. The game was played by approx. 100 k players.

6.1.3 TRUST Game

This paper presents an evaluation of the presence in a serious game that uses a virtual environment. The game studied is the TRUST game developed in conjunction with the VoiceYourView project [2]. The game's intention was to create a democratic space, in which citizens could democratically transform a virtual space (which represents Far Gosford Street, Coventry, UK) to show the local authority, planners and police, the issues which mattered most to them. The game is considered suitable to examine the relationship between presence and performance because the virtual world in the TRUST game represents a real-world space, and the actions players carry out are real-world actions, which do not require a high level of skill. Therefore, players can be expected to have comparable expectations of their performance of the tasks.

The complexity of the policy-making environment requires innovative solutions, which integrate knowledge from a range of areas, particularly from the citizens themselves. In most cases, evidence is hard data (facts, trends, survey information) and secondly the analytical reasoning that sets the hard data in context. The RCUK funded VoiceYourView [2] project recognized a gap in knowledge and its throughput into policy. Citizens, the users of public spaces have more firsthand, tacit knowledge about public spaces than officials. Yet their voices may never get heard or noticed. Citizen engagement events tend to be poorly attended or attended by an unrepresentative sample of the population. Contemporary policy articulation requires an active citizen, who is the co-designer of the policy and the main evaluator of the policy changes and decisions, and it needs to be inclusive. A cooperative virtual environment for the exchange of ideas was conceived as an easier way for citizens to articulate their concerns and display these concerns next to the location they related to.

The TRUST game idea was developed during a two-day hackathon sponsored by the VoiceYourView project. Its intention was to create democratic space, in which citizens (i.e. the users and residents of Far Gosford Street, Coventry, UK) could democratically transform a virtual space to show the local authority, planners and police, the issues which mattered most to them. In doing so, TRUST provided a mechanism for bridging the gap between citizens and local authorities.

The first stage of development was to create a virtual space that was a good representation of the real world, physical space. The second stage was to create a usable, attractive game that could be configured by users and which they had an incentive for returning to. This was done using Unity 3D [12] due to its ease of assembling different graphics modules independently, creating a collaborative environment between different developers.

In terms of awareness of visual and environmental pollution, it was easy to degrade the virtual Far Gosford Street by adding pollution and showing how easy it was to effect positive changes, e.g. by picking up litter. As the game progressed, citizens who completed the tasks were awarded credits that they could allocate to make changes to the public realm e.g. to improve street lighting or adding more rubbish bins. Those items attracting most points (from all users) would be changed in the virtual realm and an automatic request sent to the local authority requesting areas in which citizens wanted improvements to be made.

The user interface includes a functional main menu, as a starting point, and presents the environment in which the game takes place. After connection to the database, the player is introduced to the 3D representation of the game. Players can walk around the 3D world, clean graffiti, interact with game objects, links, as well as interact with other players. An overview of the user interface is shown in Figure 6.1.1.

Players were also encouraged to tailor their avatars, though in a much more basic manner. Secondly, players are given an area of the game environment in which they can tailor objects and develop them in a manner of their choosing. Finally, they are encouraged to collaborate and communicate with each other. To achieve this, a chat mechanism is built into the game and posts from other players can be pinned to owners' objects in the freeform area. This allows players to like or dislike the location, style or the objects themselves. Besides, two important entities of Trust are considered to be the teleporting portals. They are outstanding objects (in the current version they comprise of sculptures) that form a pathway between the virtual and the 3D representation of the player's real world. By clicking them, the player issues

Figure 6.1.1   Screenshot from the user interface of TRUST.

the loading of the other and it is immediately carried to the other world as requested.

6.1.4 Presence

A key measure of the success of a virtual environment is presented. Presence is defined by Wiebel et al. [13] as "the subjective feeling of immersion in a virtual environment". It's roots, as a concept are usually traced back to Minksy's [14] notion of telepresence, which refers to the feeling of operating in a remote location. Ijsselsteijn et al. [15], distinguish between social presence, such as one might feel communicating in a chat room or by video conferencing, and physical presence, "the sense of being physically located somewhere". In computer games, both forms of presence may contribute to the player experience, through interaction with other remotely located players or non-player characters, and through the illusion of playing in an imaginary world or simulated space.

Presence is of interest in the study of games, in part because its impact on the quality of the user experience, and in part because of the open question of its links to other measures, in particular performance. Weibel et al. [13] report a positive relationship between presence, flow (as defined by Csikszentmihaly [16]) and enjoyment in games. Schuemie et al. [17], in their review of investigations of presence in therapeutic applications of virtual reality, note that, while presence has subjective effects, such as intensifying the emotional reaction to a virtual experience, evidence of a relationship between presence and task performance is lacking. Furthermore, Slater [18] makes the case that there are no logical grounds for presence and performance to be linked since individuals with different levels of skill perform differently in the real world.

Applying these findings to the context of serious games, we might expect that presence would heighten players' pleasure, or engagement with the learning experience, but not impact their performance on the task. Although regular game players report that immersion in virtual words is a pleasurable experience [19], the players of serious games may be irregular players or non-players. However, considerable effort is devoted to building virtual worlds for serious games. In cases such as flight simulators, it seems reasonable to assume that a high level of presence in the virtual environment will improve the transfer of learning to the real world. However, there may be other cases where lower fidelity worlds are sufficient to support learning. For this reason, we argue that the relationship between presence and task performance in serious games warrants further investigation. In particular, we examine whether specific interface factors, which contribute to presence, also contribute to performance.

The use of questionnaires to measure the subjective experience of presence is well-established. Works that have introduced measures are summarised in Schuemie et al. [17]. Slater advocates the Immersive Tendencies Questionnaire (ITQ) [18], Usoh et al. [20] address participants' sense of presence, per se. Kim & Biaoca [21] address perceptions of 'arrival' in and 'departure' from a virtual world. Witmer & Singers' [22] Presence Questionnaire (PQ) asks questions relating to Control factors, Sensory factors, Distraction factors, and Realism factors. It has been noted, Slater [18], that the questions of the PQ do not directly address presence. Instead, its focus is on factors that are believed to influence the user's perceptions of presence. As such, it is a relevant instrument for interface designers, as its results point towards elements of the interface which may be improved.

6.1.5 Performance

What constitutes the performance of game players is, on one level, dependent upon the goals and desired outcomes of the game. Connelly et al. [23] identify a number of typical learning and behavioral outcomes of serious games, which include acquisition of knowledge and skills, but also motivational outcomes and behavioral change (it is the latter type to which the TRUST game belongs). Bellotti et al. [24] review a range of methods for in-game and post-game performance assessment. However, in the TRUST game, the activities participants undertake in the virtual world, picking up rubbish, cleaning graffiti, etc., are of an everyday type, which we would expect people to be able to do in the real world. Therefore, the performance of players at the tasks is not especially relevant: we expect all the players to accomplish all the tasks.

There is another level at which to view performance, which is performed in the game. For serious games the users may be casual or infrequent gamers, for whom the approachability of games [25] is more critical than for regular gamers, who have more familiarity with game mechanics and controls. Any feeling of presence in the virtual world is liable to be undermined by a player having to give significant attention to the user interface and its controls. Therefore, a poor user experience, one which impacts the ability to operate in the virtual world, could be expected to affect players' perceptions of their own performance. It is this aspect of performance that concerns us here.

Performance was measured using the NASA Task Load Index (TLX) questionnaire, Hart and Staveland [26]. This is a well-established instrument that has been used in at least 550 published studies in the 20 years since it was developed [27]. The NASA Task Load Index assesses workload on six scales: Mental Demand (n1), Physical Demand (n2), Temporal Demand (n3), Performance (n4), Effort (n5) and Frustration (n6). The scales range from 1 to 21 where: one to five (1–5) represents a "low" rating, six to eleven (6–11) represents "normal", twelve to sixteen (12–16) represents "high", and seventeen to twenty-one (17–21) represents a "very high" rating.

The TLX measures are self-reported, subjective evaluation of performance and workload. Since the tasks the participants undertook were not complex (removing graffiti, picking up rubbish) it was expected that most participants could complete them all in the real world. Subjective reporting indicates how well the environment supports the participant in carrying out the simulated task.

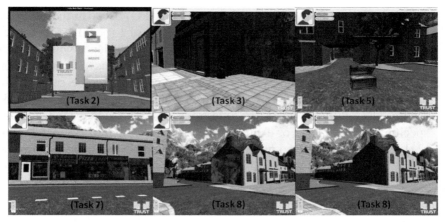

Figure 6.1.2   Overview of the tasks.

6.1.6 Method

The TRUST game was evaluated with 30 healthy participants. Demographic data collected from the participants included name, age group (minimum of 18 years), gender, current status, educational qualification, extent of daily computer usage. Participants were asked to read and understand the "Introduction to Task" instruction manual before they performed the eight tasks specified on the manual whilst thinking out loud. The tasks were 'real-life' task scenarios and the tasks were presented in numerical order. Participants were instructed to stick to the script and could ask questions where necessary. It was also clearly stated to participants that it was not obligatory to complete one task before moving on to another, but they could perform all the tasks in any numerical order, which made them feel comfortable.

The eight tasks (Figure 6.1.2) that were asked from the participants are as follows: (1) Download the Unity Web Player; (2) Start the game; (3) Navigate through Far Gosford Road and find the first bin bag; (4) Continue navigation and empty the second bin bag; (5) Locate the treasure chest and open it; (6) Locate Graffiti on Far Gosford road; (7) Locate Gogo Pizza; and finally (8) Navigate through all the way around and clean a different Graffiti on the wall.

Following completion of the tasks, two different questionnaires were given to the users. The first one aimed at testing their presence in the game while the second one measured their cognitive workload while playing the game.

The version of the Presence Questionnaire (PQ) used for this study contained twenty-six (26) questions, rated on a scale of 1 (poor) to 7 (excellent). These were used to assess the users' opinions and experience with the system they interacted with. It considers the overall reactions of participants to the system with specific emphasis on the experience of the serious game, and usability issues such ease of control. Some of the questions have been rephrased slightly to make the questions easier to understand and more specific to the games. For example, the question which in Witmer & Singer [22] reads "How involved were you in the virtual environment experience?", reads "How involved were you with the experience of the game?".

6.1.7 Results

In this section, we present summary statistics of the results of the two questionnaires.

6.1.7.1 Presence results

Figure 6.1.3 summarizes the results of the presence questionnaire PQ. The plot shows question numbers on the x-axis, with average score (orange bar) and standard deviation (whiskers). Only four of the questions had a user rating below three (3, which are questions q15, q19, q20, and q22. These questions concern delays, interference, and distraction from the task, therefore low responses are good for these questions.

As mentioned above, most of the questions of the PQ are directed at elements of the interface, which research has suggested are likely to impact upon the presence experience. However, some interrogate the presence experience more directly. The results for those questions are as follows (note that all have means above 4 and most are above 5).

- q14 "How involved were you with the experience of the game?", mean 5.2
- q21 "How completely were your senses engaged in this experience?" mean 5.53
- q24 "Were you involved in the experimental task to the extent you lost track of time" mean 4.47
- q25 "Were there moments during the experience when you felt completely focused on the task or game?" Mean 5.57

Q1. How much were you able to control events?
Q2. How responsive was the game to actions that you initiated (or performed)?
Q3. How natural did your interaction with the game seem?
Q4. How much did the visual aspects of the game involve you?
Q5. How natural was the mechanism which controlled movement of the game?
Q6. How compelling was your sense of the game moving through space?
Q7. How much did your experiences with the game seem consistent with your real world experiences?
Q8. Were you able to anticipate what would happen next in response to the action that you performed?
Q9. How completely were you able to actively survey or search the game?
Q10. How compelling was your sense of moving around the game?
Q11. How closely were you able to examine the game?
Q12. How well could you examine the game from multiple viewpoints?
Q13. How well could manipulate the game?
Q14. How involved were you with the experience of the game?
Q15. How much delay did you experience between your actions and expected outcomes?
Q16. How quickly did you adjust to the experience?
Q17. How proficient in moving and interacting with the game did you feel at the end of the experience?
Q18. How much did the visual display quality interfere or distract you from performing the assigned tasks?
Q19. How much did the control devices interfere with the performance of assigned tasks or with other activities?
Q20. How well could you concentrate on the assigned tasks?
Q21. How completely were your senses engaged in this experience?
Q22. To what extent did external events distract from your experience of the game?
Q23. Overall, how much did you focus on the game?
Q24. Were you involved in the experimental task to the extent that you lost track of time?
Q25. Were there moments during the experience when you felt completely focused on the task or game?
Q26. How easily did you adjust to the control devices used to interact with the game?

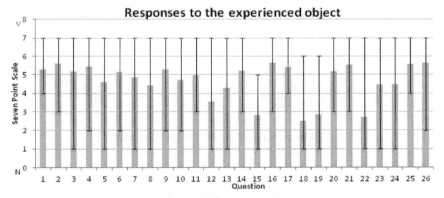

Figure 6.1.3   PQ results.

6.1.7.2 NASA TLX results

Figure 6.1.4 illustrates the results of the workload questionnaire TLX. It is notable that in response to the performance question *"How successful were you in accomplishing what you were asked to do?"* sixty-six percent (66%) of the users rated their performance as low (1–5) range, ten percent (10%) of users rated their performance as normal (6–11), seventeen percent as high (12–16) and the remaining seven percent (7%) as very high (17–21).

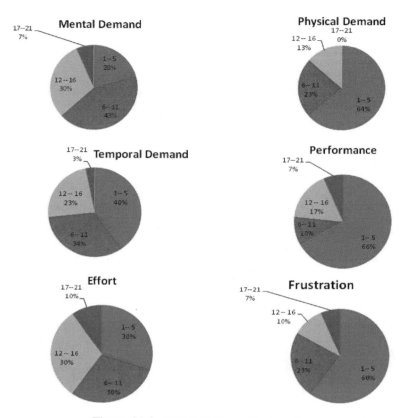

Figure 6.1.4   NASA TLX graphical results.

Twenty per cent (20%) of the 30 users who tested the game reported that the serious game had a low mental demand, forty-three percent (43%) of the users said the game required a normal mental demand, thirty percent (30%) of the users tested said the game had a high mental demand and seven percent (7%) of the users claimed that the game's mental demand was very high.

The next question related to how physically demanding the tasks were for the participants. Sixty-four percent (64%) of the participants showed that the game had low physical demand, twenty-three percent (23%) of the participants said the game had normal physical demand, thirteen percent of the users said the game had high physical demand and zero percent (0%) i.e. none of the participants, said the game had a very high physical demand.

Next, an assessed measurement of how hurried or rushed the paces of the tasks were for the participants was quantified. Forty percent (40%) of the participants mentioned that the game had a low temporal demand, thirty-four percent (34%) of the users that the temporal demand of the game task was normal, twenty-three percent (23%) of the users that the game had a high temporal demand and only three percent (3%) of the users said the task had a very high mental demand.

'Effort', aimed at measuring the amount of hard-work the participants had to do to achieve their performance level. Thirty percent (30%) of the users said that the effort required for the task was low, another thirty percent (30%) claimed that the effort required to accomplish the performance level of the task was normal. Similarly, yet another thirty percent (30%) of the users said that the effort required for the task was high and ten percent (10%) of the users said that the effort required for the tasks was very high. The final question, 'frustration' aimed at assessing how insecure, discouraged, irritated, stressed and annoyed the participants were whilst performing the tasks. Sixty percent (60%) of the users said that the frustration level was low, twenty-three (23%) of the users said that the frustration level of the tasks was normal, ten percent (10%) of the users said that the frustration level of the game tasks was high and seven percent (7%) of the participants expressed a very high frustration level.

6.1.8 Analysis

The majority of responses to the PQ suggested that the elements of the interface work well and would be expected to contribute to an experience of presence. This is supported by the positive responses to questions which more directly interrogate the experience of presence. The TLX results, however, demonstrate that the participants' subjective assessment of their own performance is not high, with only twenty-four percent (24%) of users responding with a high or very high rating. In this section, we analyze the results in more detail, looking for patterns in the responses to questions that can refine our understanding of whether specific interface factors, which contribute to presence, also influence performance. The analytical methods used are a correlation matrix, used to analyze all the responses, and parallel coordinates used on responses to questions identified from the correlation matrix for further visualization.

6.1.8.1 Correlation

A correlation matrix (r) was produced between user responses to the questions. All the questions of both questionnaires were included. The correlations are presented visually (Figure 6.1.5) using a node/edge graph layout, with questions represented as nodes and r values as edges. The display was produced using Cytoscape 3.4.0 [27] and uses Prefuse force-directed layout. Thickness of edges denotes the strength of the correlation between responses (strong red $r > 0.595$, medium $0.495 < r < 0.595$, weak $0.40 < r < 0.495$) dashed lines denote a negative value of r with strength represented by thickness in the same way. Blue nodes are the PQ questions (q1–q26). Green nodes are the TLX questions (n1–n6).

Most of the PQ questions are positively correlated to at least two other PQ questions forming a densely connected network. Of the four questions which had low responses (mean below 3), three of the questions (q15, q19, q22) sit on the periphery of the network. Of these q19 and q22 form a pair with a strong correlation to each other. The remaining question, q20 (*"How well could you concentrate on the assigned tasks or required activities, rather*

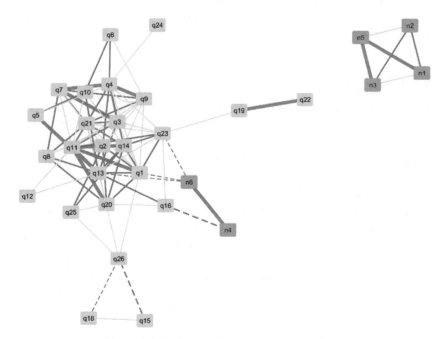

Figure 6.1.5 All correlations between questions.

than on the mechanisms used to perform those task or activities?"), forms part of a subnetwork of six nodes (q1, q5, q11, q13, q14, q20) which address aspects of control and the ability to examine and manipulate the game. The second subnetwork of strongly correlated responses (q3, q4, q7, q9) concerns naturalness, consistency with the real world, and visual aspects of the game.

Four of six of the NASA TLX questions form a disconnected clique. Within this clique Effort (n5) has strong correlations to Mental Demand (n1) and Temporal Demand (n3). This conforms with wider experience of the TLX, where the correlation between the TLX measures is commonly reported [27]. Intuitively, a connection between demand and effort seems reasonable. Disconnected from this group, Frustration (n6) and Performance (n4) form a strongly correlated pair. This is counter-intuitive, as higher levels of performance are accompanying higher levels of frustration, and vice versa. We note that both these were rated low by the majority of participants, therefore it is not that participants are typically highly frustrated but performing well. Rather they are relaxed and low performing.

There are only four correlations between the PQ questions and TLX questions. All of them are negative correlations and none of them are strong correlations. Performance (n4) is negatively correlated to one PQ question, which is q16 *"How quickly did you adjust to the experience?"*. Participants who rated their performance high (the minority) tended to report that they adjusted slowly to the experience (the minority) and vice versa. Frustration (n6) is weakly negatively correlated to three PQ questions: q1, q13, q23. Both q1 and q13 are in the previously identified subnetwork of strongly correlated questions concerning control and manipulation of the game: participants who found the game hard to control or manipulate experienced Frustration. Frustration is also negatively correlated to q23 *"Overall how much did you focus on the game?"*: participants who had low levels of frustration were able to focus.

From this analysis we identify one region of interest (ROI) which demands more detailed investigation. This is the part of the correlation matrix which has correlations between NASA TLX and PQ questions, namely the nodes n4, n6, q1, q13, q16, and q23. This is the part of the analysis which links participants' perceptions of their performance to the game design and its ability to create Presence.

6.1.8.2 Parallel coordinates

Parallel coordinates [28] stack multiple vertical axes, each representing a different dimension of the dataset, in this case a question. Each participant's responses are represented as polyline which intersects each axis at the value of that participant's response. Vertical axes can be added, removed and reordered from the analysis in the analytical package (we used HighD https://www.high-d.com/). The overlay of multiple participants' responses across the dimensions produces visual patterns, which allow us to build a more nuanced view of the data that can be gained from the correlations alone. We use it here to explore the responses in the ROI identified above. Figures 6.1.6 and 6.1.7 show the same pattern of polylines, but each is filtered in a different way, according to responses to the TLX questions, to show subsets of polylines. An axis shows the full range of the relevant scale (1–7 for PQ, 1–21 for TLX).

We know from the correlation analysis that Performance (n4) has a strong negative correlation to q16 (How quickly did you adjust to the experience?). The parallel coordinates visualization (Figure 6.1.6) gives a more detailed view of this. The upper chart shows the plot lines for participants who rated their performance above the midpoint of the scale, (24% of responses). While all the responses are in the upper half of the scale for q16 (indicating no issue with speed of adjustment), the lines form a clear crossover pattern. This would imply that the longer they took to adjust the better players rated their performance. The lower plot, filtered to show for the participants who rated their performance below the midpoint, shows no crossover pattern, but we can see that players who rated their performance as low report speed of adjustment in the same range as those who rate their performance as high.

6.1.9 Discussion and Conclusions

This study had inherent limitations, chiefly that only one game was evaluated with 30 participants. However, the results were evaluated in a systematic way using two methods of analysis to probe the data. In this way, we identified links between interface factors that affect the presence and the Performance and Frustration factors of the NASA TLX.

Performance and speed of adjustment are negatively correlated, as those participants who perceived they performed well at the TRUST game, reported taking longer to adjust to the experience. However, the more detailed visual analysis suggests that the strong negative correlation may simply result from

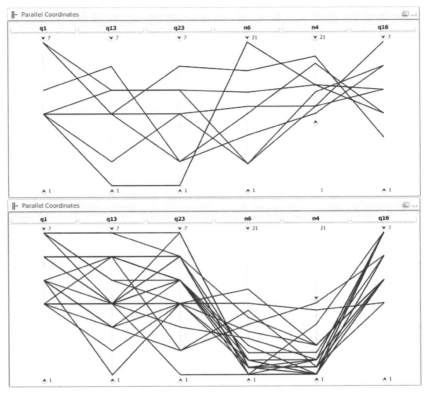

Figure 6.1.6 Parallel coordinates plots, above filtered for high responses on performance (n4 = 11−21), below low (n4 = 1−10).

the fact that the speed of adjustment of most participants is in the upper half of the range, whether they rated Performance as high or not.

On the other hand, the strong correlation between Performance and Frustration was supported by both the analytical approaches. This identifies high levels of Frustration as a partner, and potentially a cause, of good perceived Performance. The correlates of Frustration found in this study were difficulties in manipulating the game, and inability to focus on the game. These are assumed to operate independently of each other, given they are only weakly correlated to each other.

This is counter-intuitive: should developers stop trying to help inexperienced and infrequent players, and instead make games hard to manipulate and distracting? There is an alternative hypothesis: if a player is highly motivated

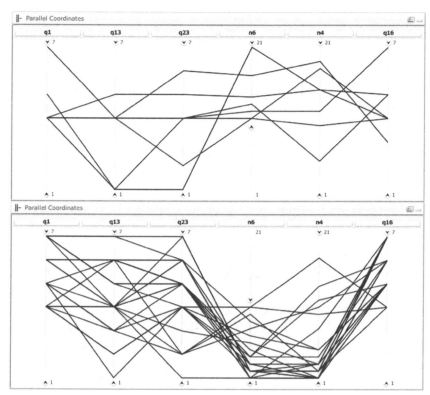

Figure 6.1.7   Parallel coordinates plots, above filtered for high responses on Frustration (n6 = 11−21), below low (n6 = 1−10).

to perform well, they may strive harder and be more inclined to get frustrated. A player who is comfortable to only achieve moderate performance may try less hard, and so experience less frustration from imperfections in the game controls. We cannot test this hypothesis with the data we have available from this study but would advocate including questions about motivation and attitude in future studies.

As serious games that employ virtual environments become more important in training and education, they will increasingly be used by players who do not regularly play such games for entertainment. However, the counter-intuitive linkage between Performance and Frustration identified in this study highlights the importance of user experience design within serious games. Helping inexperienced and infrequent players to manipulate the environment in the game and to remain focused is an essential part of the game developers' task.

Acknowledgments

A part of this work was sponsored by the VoiceYourView (vYv) research project funded by the Engineering and Physical Science Research Council's Digital Economy Program, grant number EP/H007237/1. VoiceYourView was a collaboration between five leading universities in the UK: Lancaster, Brunel, Sheffield, Manchester and Coventry University. The authors would like to thank Radu Mihaiu, Robert Eles, and Athanasios Vourvopoulos.

References

[1] Petridis, P. et al. (2015). State of the art in Business Games. *International Journal of Serious Games*, vol. 2, no. 1.

[2] Lam, B. (2013). Community-led design through digital games. *DMI Rev.*, vol. 24, no. 1, pp. 20–27.

[3] Zyda, M. (2005). From visual simulation to virtual reality to games. *IEEE Comput.*, vol. 38, no. 9, pp. 25–32.

[4] Bedwell, W. L., Pavlas, D., Heyne, K., Lazzara, E. H. and Salas, E. (2012). Toward a taxonomy linking game attributes to learning: An empirical study. *Simul. Gaming*, vol. 43, pp. 729–760.

[5] Mainemelis, C. (2001). When the muse takes it all: A model for the experience of timelessness in organizations. *Acad. Manag. Rev.*, vol. 26, pp. 548–565.

[6] Deci, E. L. and Ryan, R. M. (2004). *Handbook of Self-determination Research*. University of Rochester Press.

[7] Werbach, K. and Hunter, D. (2012). *For the Win: How Game Thinking Can Revolutionize Your Business*. Wharton Digital Press.

[8] Alhadef, E. (2012). Earth Day: Save the Environment One Facebook Serious Game at a Time with Ecotopia. *Serious Games Market*, 2012. [Online]. Available: http://seriousgamesmarket.blogspot.com/2011/04/earth-day-saveenvironment-one-facebook.html. [Accessed: 20-Jul-2017].

[9] Gamespot, "Anno 2070," *Gamespot*, 2017. [Online]. Available: https://www.gamespot.com/anno-2070/. [Accessed: 20-Jul-2017].

[10] Petridis, P. et al. (2011). "Building Social Communities around Alternate Reality Games," in *Third International Conference on Games and Virtual Worlds for Serious Applications*, 2011, pp. 76–83.

[11] Dunwell, I. et al. (2014). A game-based learning approach to road safety: The code of Everand. in *CHI '14 Proceedings of the SIGCHI Conference on Human Factors in Computing Systems*, pp. 3389–3398.

[12] Unity, "Unity 3D." [Online]. Available: https://unity3d.com/. [Accessed: 20-Jul-2017].

[13] Weibel, D., Wissmath, B., Habegger, S., Steiner, Y. and Groner, R. (2008). Playing online games against computer- vs. human-controlled opponents: Effects on presence, flow, and enjoyment. *Comput. Human Behav.*, vol. 24, pp. 2274–2291.

[14] Minsky, M. (1980). Telepresence. *Omni*, vol. 2, no. 9, pp. 45–51.

[15] IJsselsteijn, W. A., de Ridder, H., Freeman, J. and Avons, S. E. (2000). Presence: concept, determinants, and measurement. *Electron. Imaging*, vol. June, pp. 520–529.

[16] Csikszentmihalyi, M. (1975). *Beyond Boredom and Anxiety: Enjoyment and Intrinsic Motivation*. London: Jossey-Bass Inc.

[17] Schuemie, M. J., van der Straaten, P., Krijn, M. and van der Mast, C. A. (2001). Research on Presence in Virtual Reality: A Survey. *Cyberpsychology Behav.*, vol. 4, no. 2, pp. 183–201.

[18] Slater, M. (1999). Measuring Presence: A Response to the Witmer and Singer Presence Questionnaire. *Presence*, vol. 8, no. 5, pp. 560–565.

[19] Poels, K., de Kort, Y. and Ijsselsteijn, W. (2007). "It is always a lot of fun!" Exploring Dimensions of Digital Game Experience using Focus Group Methodology in *FuturePlay*, 2007, pp. 83–89.

[20] Usoh, M., Catena, E., Arman, S. and Slater, M. (2000). Using presence questionnaires in reality. *Presence*, vol. 9, no. 5, pp. 497–503.

[21] Kim, T. and Bioca, F. (1997). Telepresence via Television: Two Dimensions of Telepresence May Have Different Connections to Memory and Persuasion. *J. Comput. Commun.*, vol. 3, no. 2.

[22] Witmer, B. G. and Singer, M. J. (1998). Measuring Presence in Virtual Environments: A Presence Questionnaire. *Presence*, vol. 7, no. 3, pp. 225–240.

[23] Connolloy, T. M., Boyle, E. A., MacArthur, E., Hainey, T. and Boyle, J. M. (2012). A systematic review of empirical evidence on computer games and serious games. *Comput. Educ.*, vol. 59, pp. 661–686.

[24] Bellotti, F., Kapralos, B., Lee, K., Moreno-Ger, P. and Berta, R. (2013). Assessment in and of serious games: an overview. *Adv. Human-Computer Interact.*, p. Article ID 136864, 11 pages.

[25] Desurvive, H. and Wiberg, C. (2010). User experience design for inexperienced gamers: GAP – Game Approachability Principles. in *Evaluating User Experience in Games*, R. Bernhaupt, Ed. London: Springer-Verlag, pp. 131–147.

[26] Hart, S. G. and Staveland, L. E. (1998). "Development of NASA-TLX (Task Load Index): Results of Empirical and Theoretical Research," *Adv. Psychol.*, vol. 52, pp. 139–183.

[27] Hart, S. G. (2006). NASA Task Load Index (TLX) 20 years later. *Proc. Hum. Factors Ergon. Soc. Annu. Meet.*, vol. 50, no. 9.

[28] Inselberg, A. (2009). *Parallel Coordinates: Visual Multidimensional Geometry and Its Applications*. Springer-Verlag New York, Inc.

6.2 The Portable VR4VR: A Virtual Reality System for Vocational Rehabilitation

Rubein Shaikh[1], Paul Mattioli[1], Katey Corbett[1], Lila Bozgeyikli[2], Evren Bozgeyikli[2,*] and Redwan Alqasemi[1]

[1]Center for Assistive, Rehabilitation and Robotics Technologies, University of South Florida, USA
[2]School of Information, University of Arizona, USA
E-mail: rboz@email.arizona.edu
*Corresponding Author

Virtual reality can provide a safe and consequence-free environment to train persons with cognitive disabilities on job-related tasks. This paper is an extension of the "Virtual Reality for Vocational Rehabilitation" (VR4VR) system described in [1]. VR4VR is a virtual reality system that aims to train and assess individuals with disabilities on vocational skills in several virtual environments. It has a stationary setup which requires a few hardware items that can be considered costly for common use and requires dedicated space for permanent placement of the hardware. It is always a challenge for individuals who live far from the facility to use such a system when needed. To overcome these drawbacks, we have developed a "Portable VR4VR" system that serves the same purpose as the VR4VR but can be carried and deployed quickly in a small space of 10 by 10 feet. The new portable system is lower in cost and has minimal hardware requirements, which can be easily made available. This increases the availability and access to the system. In this paper, we present the Portable VR4VR system and present new testing results that include assessment scores of a cohort of 10 individuals with Autism Spectrum Disorder (ASD) and training scores of an individual with ASD. individuals with different disabilities. Although comparative statistical analyses are needed for robust conclusions, we find the results, comments of the participants and the vocational counselors who accompany participants throughout the user study sessions promising in the new Portable VR4VR's effective use in vocational training of individuals with ASD.

6.2.1 Introduction

About 50 million people in the United States (U.S) live with some form of disability [2]. The disability statistics in the U.S for 2016 states that

12.8% of the base population (319,215,200) have some form of disability [3]. Another report shows that out of 20,886,200 individuals who were surveyed, only 36.1% of the population with disabilities in the age group of 18 to 64 were employed [4]. Having a long or short-term disability or condition makes it difficult for individuals to get opportunities for employment, as they may find menial tasks challenging, such as socializing, navigating safely in an environment, managing time and organizing things. On the other hand, these challenges may create a prejudice on the employers as they may have a biased mindset or hesitations about employing individuals with disabilities. Nonetheless, as they were surveyed, over 66% of unemployed individuals with disabilities stated that they would like to be working [5]. Being employed is important since it was shown to provide a sense of self-achievement and economic independence in life [6, 7]. We believe this calls for endeavors in helping to create more employment opportunities for individuals with disabilities.

Training may play an important role in eliminating several barriers, such as prejudice in employers, safety concerns, time and travel-related costs. This work aims to utilize virtual reality to provide individuals with disabilities easy access to high-quality job assessment and training, and ultimately help more individuals with severe disabilities get increased employment opportunities. The main advantage of the VR4VR system is that it offers individuals with disabilities training in a controlled environment. This means that all hazard risks associated with real job places such as moving mechanical parts and sharp objects are eliminated. As the new version of the VR4VR system is portable, vocational counselors and job coaches can train and assess their clients on their premises. The portable system is compact, it can be transported anywhere and used in any room with an area larger than 10 feet by 10 feet. Individuals can be trained on different skills that are presented in Section 3: shelving boxes, money management, social skills, job interview preparation, cleaning, and environmental awareness. The variety of skills helps job coaches in determining the most suitable job for individuals by assessing them in different virtual environments.

The reason for using virtual reality (VR) is that it allows accurate replications of work environments and it offers immersive experiences with high fidelity visualization and interaction. Our intention is not to replace real-life vocational training with VR training. After individuals complete the VR training, it would still be beneficial for them to get real-life vocational training. However, as they will have then gained the basic skills within the VR system in a controlled and safe environment, we believe that the following real-life vocational training sessions would be more beneficial and smoother for them.

6.2.2 Related Work

Although there are several studies on VR, few of them focus on using VR for training individuals with disabilities. A commercial example is Specular Theory, which is a VR company that developed application that gives virtual water surfing tours to individuals with disabilities [8]. Users of the system mentioned that it was one of their best experiences as they could not have experienced it in real life. Another example is the Astro Jumper, which is a VR system that aims to encourage children with autism to exercise more [9]. Although this previous work provides motivation for designing beneficial virtual experiences for individuals with disabilities, its focus is narrowed down to exergaming and children with autism.

On the vocational side, Horace et al. developed a VR enabled approach to enhance the emotional and social adaptation skills of children with ASD [10]. The work portrayed an innovative use of virtual reality for school-aged children with ASD to assess behavioral changes through psychoeducation. The children were taught basic tasks such as washing hands, brushing teeth, etc. The authors used a four-sided Cave Automatic Virtual Environment (CAVE) to help immersing users in virtual environments. The results showed improvement in emotional recognition and social-emotional reciprocity of the participants. Mechling and Ortega-Hurndon utilized computer-based video instructions for teaching job tasks to individuals with intellectual disabilities [11]. Although it is a conventional and effective way of displaying instructions to workers, this method lacks interactivity. Another motivation behind the Portable VR4VR system gives users opportunities to have hands-on experience performing a wide range of different tasks within realistic scenarios in virtual environments.

Taking these into consideration, the VR4VR [1] was developed as an immersive 3D VR system for vocational training of individuals with disabilities. The system had novel components, such as tangible objects that were mapped into virtual worlds in real-time, with the aim of providing better user experience through improved interaction and immersion. Developing a portable version of the VR4VR system was mainly motivated by the following two factors: (1) increased accessibility and convenience for individuals with disabilities and vocational counselors and job coaches, (2) reduced prices for high accuracy VR systems.

6.2.3 Design Aspects of the Portable VR4VR

There are several aspects that need to be considered while designing a VR application for training. A few of these aspects are the complexity of the tasks, interaction methods, directions, and scenarios. The Portable VR4VR system provides audio and visual prompts and instructions similar to the original VR4VR system [1]. Visual information is known to provide a better understanding of individuals who are on the autism spectrum as compared to audio or verbal based instructions [14].

The modules in the Portable VR4VR system were designed similar to a VR video game. Complexity levels of the tasks increase gradually throughout the modules. There is also a limit the advanced difficulty stages of some of the modules. If the user fails to complete a level in a module, the following more difficult modules are not presented to them. The Portable VR4VR system includes a variety of virtual environments e.g. virtual warehouse, shopping mall, interview room, storage room, outdoor parking lot, and a cashier register in a grocery store. These virtual environments were designed to resemble real-world environments, to make the transition of the user from the virtual world to the real world easier.

The Portable VR4VR system offers seven vocational skill modules. Each module aims to train individuals on a different skill set and each module takes place in a different virtual environment. Each module is composed of three sub-modules with increasing difficulty levels. Each sub-module includes three levels (tutorial, level without distracters, level with distracters), also with increasing difficulty level for a balanced user experience.

6.2.3.1 Shelving skill module

This module involves training individuals on managing inventory, organizing shelves and managing shipping orders. There are different levels of each subtask which has different objectives i.e. sorting the boxes, order fulfillment and organizing boxes according to labels.

6.2.3.2 Environmental awareness skill module

This module involves training individuals on cautiously walking and navigating inside a virtual parking lot. The individual walks by performing a walking-in place gesture by making a marching motion. The user's movements are captured by the tracking sensors and transferred to the virtual world. Practicing this skill also helps the individual to train on motor skills.

6.2.3.3 Money management skill module

This module involves training in a virtual grocery store environment where the individual works as a cashier providing service to virtual customers. The objective of this skill is to train individuals on basic monetary mathematical skills so that they can handle financial transactions precisely in daily life. The vocational counselors and the participants we have worked with stated that the Money Management module was especially beneficial for them.

6.2.3.4 Cleaning skill module

In this module, individuals are trained or assessed in a virtual warehouse environment. The individual collects all the dirt or trash present in the virtual environment by bending and touching them. As the individual touches the dirt by bending down physically, the object gets snapped to their hands and it gets released when they stretch their hand over a virtual trash bin.

6.2.3.5 Loading the back of a truck skill module

In this module, individuals are trained in an outdoor virtual environment resembling a roadside. This module aims to train users on time and space-related problem-solving skills. The user uses a haptic device that simulates weights of objects e.g. if the box is labeled heavy then the stick of the haptic device becomes heavy and requires more pressure to be moved.

6.2.3.6 Social skills skill module

In this module, individuals get trained on social skills in different virtual environments. This module aims to train individuals on maintaining the conversation, focusing on what others say and giving relevant responses. Virtual characters ask various questions to individuals ranging from the weather conditions of that day to their opinions on how to improve a certain environment or how they would cope with an angry manager and so on.

6.2.3.7 Job interview skill module

This module aims to train individuals on job interview skills in a virtual office environment. Individuals are asked basic questions that prepare them for job interviews. If the user gives a wrong answer, the right answers are displayed on the screen, which can be observed by the individual.

6.2.3.8 Distracters

Virtual distracters were included in the third level of each skill module in the VR4VR. These distracters are in different forms such as audio, visual and animation. The aim of the inclusion of these distracters was to replicate common distractions in real workplaces and explore whether or not it would be possible to train individuals with cognitive disabilities to overcome them. In our previous user studies, we did not find statistically significant differences in user performance in levels with distracters compared to levels without distracters [1]. However, we have received positive feedback from job trainers indicating that these distractors such as a ringing phone, a thunderstorm sound or a running and screaming child help participants to get accustomed to them. In a similar way, job trainers mentioned that these distracters establish opportunities for them to observe user behavior, which can be difficult to trigger in real workplaces for training. Hence, the levels with distracters were included in the updated version of the VR4VR system.

6.2.4 Technical Aspects of the Portable VR4VR

Similar to the stationary VR4VR system in [1], the Portable VR4VR system is also designed to be responsive to the varying needs of multiple disability groups. This section describes the technical components of the Portable VR4VR system, such as hardware components and implementation.

6.2.4.1 System components

The Portable VR4VR system has four main components: motion tracking, display, software, and virtual assistive robot. These are described in detail below.

Motion Tracking: To track movement of individuals in the virtual environment, we use HTC Vive wireless trackers. Markers are attached to each hand and each foot of the individual. They are also attached to the tangible objects such as cardboard boxes and the stick which appears as the vacuum cleaner/mop in the virtual world in the Cleaning module. These markers are continuously tracked by the two HTC Vive base stations.

Display: An HTC Vive Head Mounted Display (HMD) is used to give the user an immersive experience and accurate tracking in the virtual world. In case the user gets nauseous or uncomfortable, a large 180-degree curved screen is used. This curtain screen is also used for outside viewing purposes such as displaying what a user sees inside the HMD to the vocational counselors or accompanying job coaches.

Software: The simulated virtual environments were created in the Unity game engine. The Steam kit was used for making the software compatible with the HTC Vive virtual reality headset. Scripting was performed in C#.

Seamless integration between these platforms was the major deciding factor in the selection of these platforms.

Virtual Assistive Robot: Object manipulation in virtual environment helps individuals to get a better understanding of the actual work that is done in a real-life work environment, as they can see functionality, face complications or severe problematic scenarios beforehand in a controlled environment. The Baxter robot [12] and a wheelchair mounted robotic arm [13] were used in the VR system with the aim of training individuals with physical disabilities on controlling physical robots in real-world. The virtual Baxter robot offers users practice controlling complex movements, such as carrying boxes and making a bed, more efficiently with minimal risk. This component of the VR4VR system is only being used for the assessment and training of individuals with severe physical disabilities.

A general view of the Portable VR4VR system with HTC Vive motion trackers and HMD can be seen in Figure 6.2.1. In the figure, the user wears an HMD to see the virtual environment. Motion tracking with wireless Vive trackers allows the user to interact with tangible objects in the virtual world using intuitive movements.

6.2.4.2 Implementation

The Portable VR4VR system is currently in final stage of development. There are seven modules (three with HMD and four without HMD) that were developed for the portable system. There is also an application that runs on a desktop computer independently. This software offers four modules that can be run offline. These seven modules that work both on the original VR4VR system and the new Portable VR4VR system were described in detail in [1].

To give users a more immersive experience, Vive trackers were attached to real boxes. Different textures are projected on these boxes in the virtual world. One to one motion mapping is applied to the virtual and tangible objects so that users can interact with tangible objects while they are wearing an HMD. Figure 6.2.2 shows the shelving module, in which the individual fulfills tasks by placing boxes on the correct levels of the shelves. This form of tangible interaction in virtual environments is known to increase immersion and improve user experience in VR. In Figure 6.2.3, the user mops the floor using a stick, which is rendered as a virtual mop in the virtual environment. All the movements in the real world are tracked by the motion trackers and they are transformed into the virtual environment in three-dimensions.

Figure 6.2.1 The user practices shelving tasks in a warehouse environment. The user views the virtual world through the HMD. The real-world environment is transferred to the virtual environment accurately. Different textures are projected on tangible objects.

6.2.5 User Study

In this section, we present our recent findings with the Portable VR4VR system through user studies of assessment and training, which took place between December 2018 and April 2018. The VR4VR system can be used for assessment (completion of each level of each subskill of each skill once) or training (multiple completions for reinforcement of the targeted skills). A job trainer or vocational counselor accompanies the participant during

Figure 6.2.2   Left: The user interacts with tangible boxes that are equipped with markers while seeing identical motions in the virtual world through the HMD. Right: The user interacts with tangible boxes in the shelving module. The user views the virtual environment inside the HMD. The user's view is displayed on a curtain screen for outside viewing purposes.

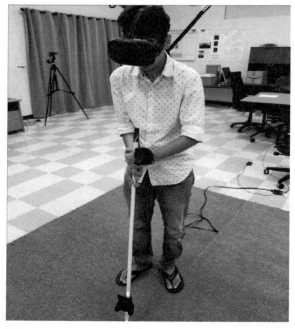

Figure 6.2.3   The user holds a tangible stick, rendered as a virtual mop. The motion of the stick is tracked by the detector.

assessment or testing. These professionals decide on the training program or the ordering of skill modules in the assessment sessions.

The participant cohort of the assessment user study consisted of 10 individuals with Autism Spectrum Disorder, who were diagnosed by medical professionals beforehand. Users were aged between 18 and 57 with $\mu = 22.3$ and $\sigma = 12.38$. Gender distribution consists of 80% males and 20 % females. These individuals were assessed using the VR4VR system with the research team and an accompanying professional job trainer or vocational counselor. The participants completed all levels of each subskill of six skill modules once and data was collected in the background. Only the following six skill modules were included in the user study: Cleaning, Loading the Back of a Truck, Money Management, Shelving, Environmental Awareness, and Social Skills. Job trainers or vocational counselors decided on the order of the skills to be assessed. The success of a participant was measured through level score and observations of the professional job trainer or vocational counselor. The level score consisted of a custom algorithm that took into account the errors and the number of prompts presented to the user, which were dependent on the time spent in a level without any success instances. The level scores were calculated out of 100. Deductions were made for the errors and prompts: -30 points for errors and -20 points for prompts, if any. The possible errors were as follows: misplacement of boxes in the shelving skill module, releasing boxes from a height in the loading the back of a truck skill module, entering wrong amounts in the money management skill module, colliding into objects, people or cars in the environmental awareness skill module and giving answers that are irrelevant to the asked questions in the social skills module. The following equation denotes the calculation of the score, where α represents the number of errors and β represents the prompts triggered.

$$\text{Level Score} = 100 - (\alpha(30) + \beta(20))$$

In Figure 6.2.4, average level scores of individuals with ASD are presented for different skills for levels without distracters, levels with distracters and the overall score, which is the average of the score in the levels without distracters and the levels with distracters for the corresponding skill. It can be observed from the chart that scores in the levels without distracters and with distracters are not considerably different, which aligns with our previous findings in [1]. To explore the overall data further, means, standard deviations and the standard error of the means are presented in Table 6.2.1. The participants received the highest scores in the Social Skills and the Cleaning skill module whereas they received the lowest scores in the Loading and

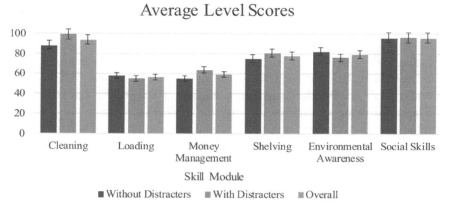

Figure 6.2.4   Average level scores of individuals with ASD in levels without distracters, levels with distracters and overall.

Table 6.2.1   Means, standard deviations and errors of the mean for each skill module's scores

Skill Module	Mean	σ	SEM
Cleaning	93.68	22.84	5.24
Loading	56.68	45.70	7.05
Money Management	59.44	46.18	5.87
Shelving	78.21	30.08	4.02
Environmental Awareness	79.77	35.00	7.46
Social Skills	96.42	6.67	0.82

the Money Management skill modules. A common characteristic of Social Skills and Cleaning Skill modules is that the tasks in these modules are very straightforward and isolated, not following progression and not requiring a lot of cognitive effort. On the contrary, Loading and Money Management skills require cognitive effort in solving analytical problems such as fitting boxes of different dimensions in a limited space or counting money.

Although comparative user studies are needed to make reliable conclusions, based on the scores following a similar trend with our previous findings in [1] and all scores' being above average (50%), we interpret that the accessible and effective design of the VR4VR system provides promising and convenient assessment and training opportunities to individuals with disabilities.

The second user study case includes training individuals with the VR4VR System. One participant with ASD was trained on different skill modules for five weeks on the following skill modules: Cleaning, Money Management,

Average Scores (Without Distracters)

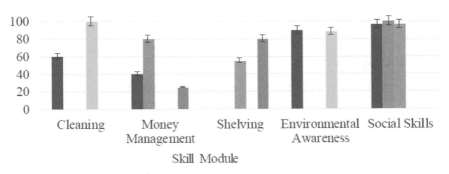

Average Scores (With Distracters)

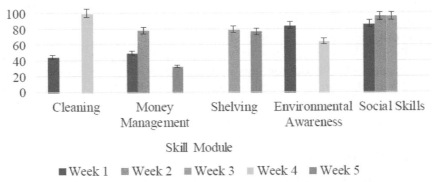

■ Week 1   ■ Week 2   ■ Week 3   ■ Week 4   ■ Week 5

Figure 6.2.5 Average level scores of individuals with ASD in levels without distracters, levels with distracters and overall.

Shelving, Environmental Awareness and Social Skills. Average score data is presented in Figure 6.2.5 for levels without distracters and levels with distracters in subsequent weeks. Although there was not enough data to perform a statistical trend analysis (since the participant did not train on each module on each week, as per the program designed by the vocational counselor), it can be observed from the chart that score of the participant increased with training for the leaning, Shelving and Social Skills modules. On the contrary, there was some level of decrease in the scores for the Money Management (for both level types) and the Environmental Awareness (only for level with distracters) modules. The training with VR4VR is a recent endeavor. With more data, comparative user studies will be conducted in order to shed light on the effects of repetitive training on skill improvement in the target population. Currently, our interpretation is that the improvement in scores of some skills is encouraging for exploring this area more.

6.2.6 Conclusion

In this paper, a portable extension to the original VR4VR system was presented. The long-term goal is to make this affordable and Portable VR4VR system that is currently in its final stages of development accessible in various parts of the world. Results from a recent assessment-based user study with a cohort of 10 individuals with ASD were shared. A training-based user case study with an individual with ASD was also presented. Although the amount of data that was collected did not allow for comparative statistical tests or trend analysis, our indications were that the Portable VR4VR system is promising for training individuals with ASD on vocational skills. The Portable VR4VR system provides most of the distinctive features of the original VR4VR system, which was found to be effective in training individuals with disabilities on vocational skills by professional vocational counselors [1]. In addition, this updated VR4VR system is more affordable and portable. Hence, we believe that the Portable VR4VR system will offer dispersed, accessible and effective vocational training to individuals with disabilities throughout the country.

6.2.7 Future Works

Our future plans include comparing participants' performance throughout several stages of their training program i.e. initial performance versus performance after multiple training sessions. More virtual environments and virtual distractors are planned to be added to the existing system, taking into consideration the input of the professional vocational counselors. More modules are also planned to be added, that focus on a single specific disability group.

References

[1] Bozgeyikli, L., et al. (2014). VR4VR: Towards vocational rehabilitation of individuals with disabilities in immersive virtual reality environments. 2014 2nd Workshop on Virtual and Augmented Assistive Technology (VAAT), Minneapolis, MN, 2014.

[2] The National Center for Biotechnology Information. Disability ratio Online Resource for U.S. Disability Ratio. 2018. [Online]. Available: https://www.ncbi.nlm.nih.gov/pmc/articles/PMC4346217/ [Accessed: 09-May-2018].

[3] Cornell University. Disability Statistics Online Resource for U.S. Disability Statistics. 2018. [Online]. Available: http://www.disabilitystatistics.org/reports/acs.cfm?statistic=1 [Accessed: 09-May-2018].

[4] Cornell University. Disability Statistics Online Resource for U.S. Disability Statistics. 2018. [Online]. Available: http://www.disabilitystatistics.org/reports/acs.cfm?statistic=2 [Accessed: 09-May-2018].

[5] Empowerment for Americans with Disabilities: Breaking Barriers to Careers and Full Employment. Washington, DC: National Council on Disability, 2007. [On-line] Available: http://www.ncd.gov/publications/2007/Oct2007 [Accessed: 09-May-2018].

[6] Schur, L. (2002). The Difference a Job Makes: The effects of employment among people with disabilities. In *Journal of Economic Issues*, Volume 36, 339–347.

[7] Yasuda, S., Wehman, P., Targett, P., Cifu, D. X. and West, M. (2002). Return to work after spinal cord injury: a review of recent research. *NeuroRehabilitation*. 17(3):177–186.

[8] National Public Radio. "Specular Theory: Online Resource for U.S. Affordable Virtual Reality", 2018. Internet https://www.npr.org/sections/healthshots/2015/10/22/450573400/affordable-virtual-reality-opens-newworlds-for-people-with-disabilities [Accessed: 09-May-2018].

[9] Finkelstein, S. L., Nickel, A., Barnes, T. and Suma, E. A. (2010). Astrojumper: Designing a virtual reality exergame to motivate children with autism to exercise. In *Proceedings of the 2010 IEEE Virtual Reality Conference (VR '10)*. IEEE Computer Society, Washington, DC, USA, 267–268.

[10] Ip, H. H., Wong, S. W., Chan, D. F., Byrne, J., Li, C., Yuan, V. S., Lau, K. S. and Wong, J. Y. (2018). Enhance emotional and social adaptation skills for children with autism spectrum disorder: A virtual reality enabled approach. *Computers & Education*, 117, pp. 1–15.

[11] Mechling, L. C. and Ortega-Hurndon, F. (2007). Computer-based video instruction to teach young adults with moderate intellectual disabilities to perform multiple step, job tasks in a generalized setting. *Educ Train Develop Disabil*. 42:24–37.

[12] Rethink Robotics Baxter Robot Internet: http://www.rethinkrobotics.com/products/baxter/

[13] Shrock, P., Farelo, F., Alqasemi, R. and Dubey, R. (2009). Design, simulation and testing of a new modular wheelchair mounted robotic arm to perform activities of daily living. In Proceedings of the 2009 IEEE 11th International Conference on Rehabilitation Robotics. Japan, June 23–26, 2009.

[14] Teching Students with Autism Spectrum Disorders, Department of Education, New Nouveau Brunswick, 2005.

6.3 VRTouched: Towards Exploring Effects of Tactile Communication with Virtual Robots on User Experience in Virtual Reality

Evren Bozgeyikli

School of Information, University of Arizona, USA
E-mail: rboz@email.arizona.edu

In this chapter, we present a work-in-progress about examining the effects of affective touch in human-virtual robot interaction on user experience in virtual reality (VR), which is an unexplored area. We introduce related works, present the design of our system that we named '*VRTouched*', mention user study plans, discuss the challenges we have faced so far along with alternative solutions, and finally conclude with future research plans. The main motivation behind this chapter is to raise awareness in this unexplored area of tactile human-virtual robot interaction in VR, to inspire and leverage future studies in this field.

6.3.1 Introduction

Affective touch has been known for a long time to have a crucial role in human relationships, which affects several human emotions and reactions such as mood, motivation and effort facilitation [1, 2]. As it can be observed widely in daily human to human interactions in various forms such as approving, playful or controlling touches; tactile communication also became an emerging field in human-robot interaction in recent years [3]. Various studies have shown that humans react to robot-human touch similarly as they react to human-human touch [4–9]. This emphasizes the importance of touch in human-robot interaction design for an improved user experience. With the prediction of robots to be increasingly prevalent in the near future, and the significant advancements in robotics; several researchers directed towards working on making robots socially more intelligent and interactive, and have been developing novel ways of making robots closer to humans in terms of touch interaction [10–14]. These studies were in several different domains such as service [15], health care [16], companionship [17, 18] and all utilized real (tangible) robots.

There are also virtual robots that are virtual characters in the shape of robots. These virtual robots have been used in video games, education and training applications on different platforms such as computers, game consoles, and virtual reality. Virtual reality (VR) can be described as a model of reality that is composed of computer-generated environments that can be experienced by users through exploration and interaction (usually in three-dimensions) [19]. VR has been linked to several advantages such as increased safety as compared with real-world encounters which can sometimes be dangerous (e.g.: sharp objects, moving heavy machinery parts, traffic accidents), high visualization capabilities and effective training through highly customizable and automated scenarios with real-time feedback and assistive prompts [19–22].

Although there have been several researches on human-robot touch interaction, effects of tactile communication between humans and virtual robots in VR has not been explored yet. However, as VR is an emerging area that is predicted to be increasingly prevalent in the near future and it includes common use of virtual robots, the intersection of tactile communication with robots and virtual reality is an important area. The work-in-progress described in this chapter aims to address this gap.

6.3.2 Background

Although there is no study to our knowledge that explored effects of tactile communication in virtual reality robotics, several studies have examined this area with real robots. Huggable was a furry effective robot with the capability to participate in affective touch-based interactions with people and was designed as a therapy animal [23]. Cuddlebot was an animal robot that facilitated gesture aware touch interactions to reduce stress and anxiety in humans [24]. Haptic Creature was an animal robot that recognized and reacted to affective touch [25]. As the previous research in VR robotics is considered, a few studies have examined how humans respond to virtual robot characters [26–28]. Although these studies' foci were different, a common finding throughout them was that VR displays would increase the perceptual benefits in terms of virtual robots as compared to flat screens such as monitors.

6.3.3 The VRTouched System

In this study, we present a work-in-progress that aims to explore the effects of touch in human-virtual robot interaction on user experience in virtual reality.

For this, a custom VR system is currently being developed at the Extended Reality and Games Laboratory (XRG Lab) of the School of Information (iSchool) at the University of Arizona.

6.3.3.1 Design and components of the VRTouched system

The diagram sketch of the proposed VRTouched system is presented in Figure 6.3.1. An HTC Vive VR system with two motion tracking lighthouses is used. The user wears a head-mounted display for viewing the virtual world and converses with a virtual character that can be in the form of either a robot or virtual human. During these conversations, the virtual character occasionally touches the virtual avatar of the user in an affirmative way, which is accompanied by a real-time physical touch in real-world through the custom-designed tangible system. The location and the timing of the touch is synchronized with the virtual touch to increase the degree of perceived realism of the touch. This system was designed to be in the form of a wearable backpack with attached components. The backpack is worn by the user prior to starting using the VR system.

User studies will be conducted to make comparisons between users' reactions to the following: (1) virtual versus tactile communication with virtual

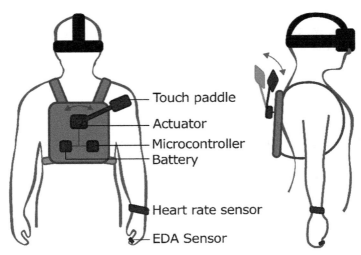

Figure 6.3.1 Sketch of the VRTouched. The user wears a backpack with custom built components attached to it. Touch paddle is rotated to touch the user whenever the virtual robot touches the virtual character. This gives the user a tactile touch feedback, accompanying the virtual touch.

human characters, (2) virtual versus tactile communication with virtual robot characters, (3) differences between reactions to tactile communication with virtual human characters versus virtual robot characters. Users' responses will be measured using Electrodermal Activity (EDA) and heart rate sensors, as these two measuring methods have been commonly used in measuring human responses to the physical touch of real-world robots [29]. Other than these, user experience metrics that are prevalently used in VR studies [19] such as level of presence and enjoyment will be collected through question-naires after the users complete the VR simulation. User studies are planned to be performed after the development of the system is completed and in-house testing is performed.

6.3.3.2 Challenges

Challenges associated with the design and development of the VRTouched system can be listed as follows: (1) a seamless tactile interface that would give a realistic touch feeling yet would not be noticed by the users; (2) synchronizing the positions of the touches in the virtual world and the real world. For the first challenge, we came up with the solution of design-ing a wearable prototype with a dynamic touch surface that is similar to a paddle. The prototype caters for different sized human bodies as the touch paddle's direction can be adjusted prior to the VR simulation. For the second challenge, we came up with the solution of getting the position data of the touch paddle and matching the position of the virtual touch with this data dynamically.

6.3.4 Conclusion and Future Work

This paper presents a work-in-progress which aims to explore effects of tactile communication with virtual robots on user experience in virtual reality. The proposed VRTouched system's design and current development sta-tus was described along with user study plans, challenges, and alternative solutions.

References

[1] J. D. Fisher, M. Rytting, and R. Heslin, "Hands Touching Hands: Affective and Evaluative Effects of an Interpersonal Touch," *Sociometry*, vol. 39, no. 4, pp. 416–421, 1976.

[2] N. Guéguen, "Touch, awareness of touch, and compliance with a request," *Perceptual and Motor Skills*, vol. 95, no. 2, pp. 355–360, 2002.

[3] E. Kerruish, "Affective Touch in Social Robots." *Transformations (14443775)*, vol. 29, 2017.

[4] R. Andreasson, B. Alenljung, E. Billing, and R. Lowe, "Affective touch in human-robot interaction: Conveying Emotion to the Nao Robot." *International Journal of Social Robotics*, pp. 1–19, 2017.

[5] T. L. Chen, C. A. King, A. L. Thomaz, and C. C. Kemp, "An investigation of responses to robot-initiated touch in a nursing context." *International Journal of Social Robotics*, vol. 6, no. 1, pp. 141–161, 2014.

[6] T. Arnold, and M. Scheutz. "Observing Robot Touch in Context: How Does Touch and Attitude Affect Perceptions of a Robot's Social Qualities?." In *Proceedings of the 2018 ACM/IEEE International Conference on Human-Robot Interaction*, ACM, pp. 352–360, 2018.

[7] J. J. Li, W. Ju, and B. Reeves, "Touching a mechanical body: tactile contact with body parts of a humanoid robot is physiologically arousing." *Journal of Human-Robot Interaction*, vol. 6, no. 3, pp. 118–130, 2017.

[8] M. Shiomi, K. Nakagawa, K. Shinozawa, R. Matsumura, H. Ishiguro, and N. Hagita, "Does a robot's touch encourage human effort?." *International Journal of Social Robotics*, vol. 9, no. 1, pp. 5–15, 2017.

[9] M. D. Cooney, S. Nishio, and H. Ishiguro, "Importance of touch for conveying affection in a multimodal interaction with a small humanoid robot." *International Journal of Humanoid Robotics*, vol. 12, no. 01: 1550002, 2015.

[10] D. Mazzei, C. De Maria, and G. Vozzi, "Touch sensor for social robots and interactive objects effective interaction." *Sensors and Actuators A: Physical*, vol. 251, pp. 92–99, 2016.

[11] T. Hirano, M. Shiomi, T. Iio, M. Kimoto, T. Nagashio, I. Tanev, K. Shimohara, and N. Hagita, "Communication cues in a human-robot touch interaction." In *Proceedings of the Fourth International Conference on Human Agent Interaction*, ACM, pp. 201–206, 2016.

[12] P. Orefice, M. Ammi, M. Hafez, and A. Tapus, "Let's handshake and i'll know who you are: Gender and personality discrimination in human-human and human-robot handshaking interaction." In *IEEE-RAS 16th International Conference on Humanoid Robots (Humanoids)*, IEEE, pp. 958–965, 2016.

[13] R. D. P. Wong, J. D. Posner, and V. J. Santos, "Flexible microfluidic normal force sensor skin for tactile feedback." *Sensors and Actuators A: Physical*, vol. 179, pp. 62–69, 2012.

[14] A. Aly, S. Griffiths, V. Nitsch, T. Taniguchi, S. Wermter, and A. Tapus, "Day Workshop Towards Intelligent Social Robots: Social Cognitive Systems in Smart Environments." *IEEE Ro-Man Portugal*, 2017.

[15] K. S. Jones, and E. A. Schmidlin, "Human-robot interaction: toward usable personal service robots." *Reviews of Human Factors and Ergonomics*, vol. 7, no. 1, pp. 100–148, 2011.

[16] E. Broadbent, R. Stafford, and B. MacDonald, "Acceptance of health-care robots for the older population: Review and future directions." *International Journal of Social Robotics*, vol. 1, no. 4, p. 319, 2009.

[17] T. Kanda, T. Hirano, D. Eaton, and H. Ishiguro, "Interactive robots as social partners and peer tutors for children: A field trial." *Human-Computer Interaction,* vol. 19, no. 1, pp. 61–84, 2004.

[18] M. M. Jung, L. van der Leij, and S. M. Kelders, "An exploration of the Benefits of an animallike robot companion with More advanced Touch interaction capabilities for Dementia care." *Frontiers in ICT*, vol. 4, p. 16, 2017.

[19] K. S. Hale, and K. M. Stanney, "Handbook of virtual environments: Design, implementation, and applications." CRC Press, 2014.

[20] A. S. Rizzo and G. J. Kim, "A SWOT analysis of the field of virtual reality rehabilitation and therapy." *Presence: Teleoperators and Virtual Environments*, vol. 14(2), pp. 119–146, 2005.

[21] C. Putnam, and L. Chong, "Software and technologies designed for people with autism: what do users want?" In *Proceedings of the 10th International ACM SIGACCESS Conference on Computers and Accessibility*, ACM, pp. 3–10, 2008.

[22] H. Sharp, Y. Rogers, and J. Preece, "Interaction design: beyond human-computer interaction" Wiley, 2007.

[23] W. D. Stiehl, C. Breazeal, K. Han, J. Lieberman, L. Lalla, A. Maymin, J. Salinas et al., "The huggable: a therapeutic robotic companion for relational, affective touch." In *ACM SIGGRAPH 2006 Emerging Technologies*, ACM, p. 15, 2006.

[24] J. Allen, L. Cang, M. Phan-Ba, A. Strang, and K. E. MacLean, "Introducing the CuddleBot: A robot that responds to touch gestures," in *Proceedings of the Tenth Annual ACM/IEEE International Conference on Human-Robot Interaction Extended Abstracts*, ACM, pp. 295–295, 2015.

[25] S. J. Yohanan, "The haptic creature: Social human-robot interaction through affective touch." PhD dissertation, University of British Columbia, 2012.

[26] O. Liu, D. Rakita, B. Mutlu, and M. Gleicher, "Understanding human-robot interaction in virtual reality." In *26th IEEE International Symposium on Robot and Human Interactive Communication (RO-MAN)*, IEEE, pp. 751–757, 2017.

[27] J. T. C. Tan, T. Inamura, K. Sugiura, T. Nagai, and H. Okada, "Human-robot interaction between virtual and real worlds: Motivation from robocup@ home." In *International Conference on Social Robotics*, Springer, pp. 239–248, Cham, 2013.

[28] B. Armstrong, D. Gronau, P. Ikonomov, A. Choudhury, and B. Aller, "Using virtual reality simulation for safe human-robot interaction." *Indiana University Purdue University Fort Wayne (IPFW) Illinois-Indiana and North Central Joint Section Conference*, 2006.

[29] M. Ménard, P. Richard, H. Hamdi, B. Daucé, and T. Yamaguchi, "Emotion recognition based on heart rate and skin conductance." In *PhyCS*, pp. 26–32, 2015.

6.4 Emerging Challenges for HCI: Enabling Effective Use of VR in Education and Training

Neil A. Gordon and Mike Brayshaw

Department of Computer Science and Technology, University of Hull, UK

This chapter considers some of the challenges in providing effective virtual reality (VR) environments for teaching and training, where users are encouraged and enabled to be truly engaged in their learning. One approach is to use inquiry-based learning, linking that to the use of VR as a vehicle for education. This chapter introduces the notion of virtual learning spaces in a more general form and makes the case that a virtual learning space is any online environment where the learner perceives that they are interacting or gaming, and thus can be enabled within a virtual environment. Thus, virtual learning spaces are not limited to bespoke education learning software but can be considered in any context where the user perceives that they are engaging, having fun, and exploring, as can be implemented within a VR platform.

Infotainment or gaming is a potential area that we can exploit to improve the effectiveness of training and learning and is particularly relevant within VR learning spaces. This chapter considers both synchronous and asynchronous computer-mediated communication and virtual and immersed virtual reality to illustrate potential virtual learning spaces. The chapter presents some worked illustrations to show how this could be used with the context of teaching Computer Science.

6.4.1 Introduction

The potential to utilize VR, augmented reality and related technology in teaching and learning is growing as the cost lowers and the availability increases. There are several emerging technologies, though these form a spectrum rather than discrete points as the facilities improve. Display technologies for VR include headsets and CAVES that immerse users in computer-generated environments. These can enable immersive training and education [1]. The availability of domestic mainstream technology means that augmented learning is viable [2]. These can be built using bespoke technologies, though many are now built using game environments, and utilizing game engines, and thus falls into the area of serious games, with learning outcomes intended as part of the virtual experience.

This area encompasses the range of technologies, from VR, virtual worlds, virtual environments, augmented reality through to mixed reality. The opportunity to present these through mobile devices, bespoke headsets, as well as interaction through more traditional 2D and 3D computer screens means the distinction between one and another is decreasing and is reflected in the umbrella term of extended reality (XR) [3].

6.4.1.1 Virtual reality and education

The growth of virtual augmented and mixed reality in education is growing, though is still in its early stages in terms of its application and evidence of its impact on learning itself [4]. As the cost of equipment is coming down, and with the opportunity for low cost virtual and augmented reality through mobile devices, the opportunity for the use of this type of equipment is expanding, and becoming more mainstream, with new build schools now routinely including virtual reality studios and capacity, whilst solutions such as Google Cardboard mean that the cost of accessing VR type equipment is no longer a significant barrier. However, the effective use of these approaches is also relatively early in its development, with numerous issues in terms of its usability [5].

VR applications in education encompass the opportunities to place learners in a variety of scenarios, and in particular to let them experience different environments whilst situated in the classroom, with tools such as Google Expeditions enabling that with relatively low-cost equipment [6]. Augmented Reality can similarly offer tools to situate computer-generated content within the physical learning space [7].

One of the challenges from a human-computer interaction (HCI) perspective, is how to effectively design and assess the design of, these learning environments, and how to effectively utilize them to truly enhance learning.

6.4.1.2 Usability and HCI evaluation

Usability testing approaches for VR and AR are improving, as the field is maturing [8, 9]. However, as learning environments, a challenge is how to effectively assess their effectiveness, both from an HCI perspective (i.e. how usable are they) and from an effectiveness as a learning environment (i.e. how far do they improve the learner's knowledge and understanding?)

Usability testing can encompass a range of approaches that can be applied to VR and AR systems. When considering their effectiveness as learning and training environments, this adds further dimensions to consider from an HCI perspective.

One approach to usability, that can allow for rapid evaluation of a range of platforms for learning, is heuristic evaluation [10], which has been adapted and used for the evaluation of learning environments [11], and more latterly adopted for AR and VR environments [12–14]. One particular aspect of HCI and usability testing for VR/AR applications, is the physical side, both in terms of the interaction, and also in terms of the impact on the user. Side effects – e.g. nausea and disorientation – in VR affect the planning of usability testing, and potentially the focus of the users [15]. Aside from the interaction itself, there is also the issue of how to assess the impact on the learner's understanding and knowledge, i.e. what is the learning gain from using such a platform?

6.4.1.3 Enquiry-based learning

Inquiry-based learning [16–19] aims to place the learner at the center of the learning process. They learn in a flexible way, at their own pace, in their own language, following their own paths of exploration and inquiry [20]. In this way, the learner's models of their domain of inquiry develop as they discover new things and expand their vocabulary. This approach to learning does not attempt to impose a set model on the learner but allows them to explore - although it has been noted that this process may need some guidance [21, 22]. It thus differs from traditional didactic approaches, as rather than the learner being a passive receiver, they have autonomy, whilst the instructor aims to provide a stimulating environment, to enable, foster, stimulate, and encourage the learner to follow their own learning route. If a learner finds the learning on their own, as an expansion of their current mental model, they will understand both the language and context of their new learning, what it means to them, and what to do with it within their current cognitive dynamics. A virtual environment – or an augmented real environment – can provide ways to let learners explore the virtual world/augmented real world, thus allowing for true and active inquiry style learning.

One of the big issues in education is motivation. How can we persuade people that learning is important and that it is something in which they should engage? For this reason, people have sought methods and techniques to encourage this participation. Now one area where we see high levels of motivation is computer gaming [23]. There has thus been a great deal of work to try and marry these levels of motivation and encouragement to learn (e.g. WEST [24], Wumpus, WUSOR-II [25–27]. The argument, in essence, is that if we transfer the levels of engagement and enthusiasm for playing computer games to learning using a computer we can better deliver education and the motivation and time spent will deliver better learning outcomes Again, this is an area where virtual reality can be exploited, particularly with 3D game engines providing the toolset to create worlds and utilize game style mechanics to motivate the learner/player.

Another area of high motivation is social computing. As people are glued to games there are also glued to social media with the likes of Facebook, Twitter, WhatsApp, YouTube channels, and other media sharing and broadcasting applications. In all of these, there is a notion of sharing something. In terms of sharing a VR this can either be explicit for example in a serious game where users are embedded in a shared and directly experienced 3D world. The aim in this type of experience is to mimic the real world as closely as possible so the result of the simulation and the type of experience you would have in the real world are a close match. The second type of VR is where we share the same metaphor of reality although it is actually far from believable. Second Life [28] is a good example of this. The culture of online Vloggers is another. Users share the metaphor that this maps onto something real although it is clearly not. There is an element of play here also, in that everyone has to indulge in shared beliefs and a made-up reality in order to make it work. Many VR learning spaces are of this second type. They need not map closely to reality, but as long as the reality metaphor holds up, or the common willingness to pretend that this is the case, then we can behave as if we are in the real thing.

6.4.2 Virtual Learning and Training Spaces

It has been argued elsewhere that humans are fundamentally game players both in nature and culture [29]. In this chapter, we will argue that what is a game is very much in the mind of the person engaging in the activity. Indeed, what is a game that may not be a game to others? Again, we can make the distinction between environments that try to directly reflect reality. For example flight simulation games whereby, they are so realistic that you can actually learn to fly, in contrast many activities may be undertaken and the participants regard that undertaking as infotainment or a game whether the designers of that activity originally intended it to be such. For example, one mental model of the auction house eBay is that of a game that can be played night after night. It clearly is not like a virtual reality auction house, but we can enter into the metaphor and it becomes a virtual one. The gamble is that you get what you think you have bought. The rewards of the game are to get a bargain, and the particular extra rewards are if you have used your skill and domain knowledge to spot something the rest of the world (including the seller) have not spotted. This activity is often called gamification. This is taking an existing activity and turning it into a game. As long as the participants are in a gaming context in their heads then it is a game to them. In this context many everyday activities can be made to be games. With virtual, augmented and extended reality, the scope to utilize an immersive and interactive experience as a way to enable inquiry-based learning, means that this can be done explicitly. A challenge here is to ensure that the intended learning is happening and that this is not lost in the experience itself.

So, a challenge is how to ensure a true virtual learning space within a virtual environment. In some response, any online area that the participants engage with computer-based collaboration can be seen as a virtual learning space. The VR learning spaces may be individual or involve others in a shared experience. If the nature of that shared interaction leads to a learning experience, then we can talk about collaborative learning in a virtual space.

Note that such learning encompasses training too – so the form of learning may be more practical and vocational, rather than academic. Indeed, in many respects, VR/AR/XR is better suited to such vocation focus. Examples of such training are more common, as the cost savings and impact on safety are typically more explicit and greater in value. This can include training fire-fighters [30], wind-turbine engineers [31], as well as pilots [32] as mentioned above.

6.4.3 Integrating XR into a VLE

Effective use of VR/AR/XR in learning and teaching means managing the HCI aspects in two distinct – yet complementary – dimensions; in one dimension, ensuring the HCI in terms of the usability of the virtual environment; in the other dimension, considering the HCI in terms of the educational payload. This assessment of the usability and effectiveness of the system is further complicated by the need to situate it alongside a wider set of tools. For example, Figure 6.4.1 outlines the way that a VR/AR/XR system would sit within a wider framework of a learning platform (VLE), and various functions that support the training/education application context. This can be considered as several layers, with the need to support the VR display

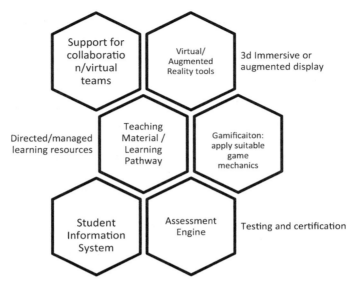

Figure 6.4.1   A framework for supporting learning in VR/AR systems. Adapted from [33].

and environment, the need to enable and encourage inquiry and exploration, and the requirement to track progress and assess the student (testing and verification).

6.4.4 Conclusions

As discussed in this chapter, inquiry-based learning aims to place the learner at the center of the learning process. By providing a flexible environment for learning, learners can study at their own pace, in their own language, following their own paths of exploration and inquiry. VR and associated technologies offer the potential to create and enable stimulating environments for learning, enabling and stimulating the learner so they can follow their own learning route. Such flexible learning – utilizing game mechanics alongside the immersive/augmented technologies now available – has significant promise in terms of new learning opportunities. Integrating this into/alongside existing training and learning platforms would enable greater use and exploitation of this opportunity. However, ensuring effective design of such systems, in terms of the user experience as both an immersive experience, and as a learning instrument, will require the development of effective HCI techniques to assess both the usability of the virtual environment, as well as the impact on the learning offered by the virtual learning environment: truly immersing VLE into VE.

References

[1] Appelman, R. (2005). Designing experiential modes: A key focus for immersive learning environments. TechTrends 49:3 pp. 64–74.

[2] Herrington, J., Reeves, T. C. and Oliver, R. (2007). Immersive learning technologies: Realism and online authentic learning. Journal of Computing in Higher Education 19(1), pp. 80–99.

[3] North of 41, (2018). What really is the difference between AR/MR/VR/XR?, Available online: https://medium.com/@northof41 /what-really-is-the-difference-between-ar-mr-vr-xr-35bed1da1a4e Accessed 10/05/2019.

[4] Hussein, M. and Nätterdal, C. (2015). The benefits of virtual reality in education-A comparison study. Ph.D. Goteborgs University.

[5] Coban, M., Karakus, T., Karaman, A., Gunay, F. and Goktas, Y. (2015). Technical problems experienced in the transformation of virtual worlds into an education environment and coping strategies. Journal of Educational Technology & Society, 18(1), pp. 37–49.

[6] Brown, A. and Green, T. (2016). Virtual reality: Low-cost tools and resources for the classroom. TechTrends, 60(5), pp. 517–519.

[7] Billinghurst, M. and Duenser, A. (2012). Augmented reality in the classroom. Computer, 45(7), pp. 56–63.

[8] Dünser, A., Grasset, R., Seichter, H. and Billinghurst, M. (2007). Applying HCI principles to AR systems design.

[9] Chang, A., Paz, F., Arenas, J. J. and Díaz, J. (2018). Augmented reality and usability best practices: A systematic literature mapping for educational videogames. In 2018 IEEE Sciences and Humanities International Research Conference (SHIRCON) (pp. 1–5). IEEE.

[10] Nielsen, J. and Molich, R. (1990). Heuristic evaluation of user interfaces. In Proceedings of the SIGCHI conference on Human factors in computing systems (pp. 249–256). ACM.

[11] Gordon, N., Brayshaw, M. and Aljaber, T. (2016). Heuristic evaluation for serious immersive games and M-instruction. In International Conference on Learning and Collaboration Technologies (pp. 310–319). Springer, Cham.

[12] de Almeida Pacheco, B., Guimarães, M., Correa, A. G. and Martins, V. F. (2018). Usability evaluation of learning objects with augmented reality for smartphones: A reinterpretation of Nielsen Heuristics. In Iberoamerican Workshop on Human-Computer Interaction (pp. 214–228). Springer, Cham.

[13] Sutcliffe, A. and Gault, B. (2004). Heuristic evaluation of virtual reality applications. Interacting with Computers, 16(4), pp. 831–849.

[14] Sutcliffe, A. G., Poullis, C., Gregoriades, A., Katsouri, I., Tzanavari, A. and Herakleous, K. (2019). Reflecting on the design process for virtual reality applications. International Journal of Human-Computer Interaction, 35(2), pp. 168–179.

[15] Arrambide, K. (2018). HCI Games Group. User Testing for Virtual Reality (VR) Headsets. Available online: https://medium.com/@hci gamesgroup/user-testing-for-virtual-reality-vr-headsets-ea5549e6f16e Accessed 10/05/2019.

[16] Vygotsky, L. S. (1934). Though and Language, Alex Kozulin (Ed), MIT Press, 1986, 0-262-72101-8.

[17] Bruner, J. S. (1961). The act of discovery. Harvard Educational Review, 31, 21–32.

[18] Bruner, J. S. (1966). Toward a theory of instruction, Cambridge, Mass.: Belkapp Press.

[19] Wood, D. J., Bruner, J. S. and Ross, G. (1976). The role of tutoring in problem-solving. Journal of Child Psychiatry and Psychology, 17(2), 89–100.

[20] Gordon, N. (2014). Flexible pedagogies: Technology-enhanced learning. York: Higher Education Academy.

[21] Elsom-Cook, M. (1984). Design considerations of an intelligent tutoring system for programming languages (Doctoral dissertation, University of Warwick). Elsom-Cook, 1990 Multimedia book.

[22] Butterfield, A. M. and Brayshaw, M. (2014). A pedagogically motived guided inquiry-based tutor for C#. In Proceedings of the HEA STEM (Computing) Learning Technologies 2014 Workshop, University of Hull.

[23] Rigby, S. and Ryan, R. M. (2011). Glued to Games: How video games draw us in and hold us spellbound, Praeger.

[24] Carr, B. P. and Goldstein, I. P. (1977). Overlays: a theory of modeling for computer-aided instruction, AI Memo 406, AI Laboratory, Massachusetts Institute of Technology.

[25] Goldstein, I. P. (1977). The computer as coach: An athletic paradigm for intellectual education, AI Memo 389, AI Laboratory, Massachusetts Institute of Technology.

[26] Goldstein, I. P. (1979). The generic epistemology of rule systems, International Journal of Man-Machine Studies, 11, pp. 51–77.

[27] Stansfield, J. L., Carr, B. P. and Goldstein, I. P. (1976). WUMPUS Advisor 1: A first implementation of a program that tutors logical and probabilistic reasoning skills, AI Memo 381, AI Laboratory, Massachusetts Institute of Technology.

[28] Rymaszewski, M., Au, W. J., Wallace, M., Winters, C., Ondrejka, C. and Batstone-Cunningham, B. (2007). Second life: The official guide. John Wiley & Sons.

[29] Huizinga, J. (1949). Homo ludens; a study of the play element in culture. Routledge & Kegan Paul Ltd.

[30] Focus, V. R. (2018). RiVR Develop New VR Solution For Fire Service Training. Available https://www.vrfocus.com/2018/06/rivr-develop-new-vr-solution-for-fire-service-training/ Accessed 4 June 2019.

[31] 4C Offshort, (2019). Virtual reality for offshore training. Available https://www.4coffshore.com/news/virtual-reality-for-offshore-training-nid11097.html Accessed 4 June 2019.

[32] Seidel, R. J. and Chatelier, P. R. eds. (2013). Virtual reality, training's future?: perspectives on virtual reality and related emerging technologies (Vol. 6). Springer Science & Business Media.

[33] Gordon N. and Brayshaw M. (2017). Flexible Virtual Environments: Gamifying Immersive Learning. In: Stephanidis C. (eds) HCI International 2017 – Posters' Extended Abstracts. HCI 2017. Communications in Computer and Information Science, vol. 714. Springer, Cham.

Index

About the Editors

Lila Bozgeyikli is an Assistant Professor at the University of Arizona's School of Information (iSchool). Her research interests include extended reality, with a focus on blurring the boundary between virtual and real worlds and application of extended reality in improving lives, such as education, training, health and well-being; video game development; human-computer interaction. She earned her PhD in Computer Science and Engineering from the University of South Florida. She has an MSc on Game Technologies.

Ren Bozgeyikli is an Assistant Professor at the University of Arizona's School of Information (iSchool). His research interests include interaction (i.e., object manipulation and locomotion) in extended reality, with the aim of improving user experience and leveraging the benefits that are gained by the society from these systems; video game development, algorithms in video games, and human-computer interaction. He earned his PhD in Computer Science and Engineering from the University of South Florida. He has an MSc on Game Technologies.